THE SPECIAL THEORY
OF RELATIVITY

THE SPECIAL THEORY
OF RELATIVITY
An Introduction

Dennis Morris

MERCURY LEARNING AND INFORMATION

Dulles, Virginia
Boston, Massachusetts
New Delhi

Publisher: David Pallai
MERCURY LEARNING AND INFORMATION
22841 Quicksilver Drive
Dulles, VA 20166
info@merclearning.com
www.merclearning.com
1-800-232-0223

This book is printed on acid-free paper.

Dennis Morris. *The Special Theory of Relativity: An Introduction.*
ISBN: 978-1-942270-72-0

The publisher recognizes and respects all marks used by companies, manufacturers, and developers as a means to distinguish their products. All brand names and product names mentioned in this book are trademarks or service marks of their respective companies. Any omission or misuse (of any kind) of service marks or trademarks, etc. is not an attempt to infringe on the property of others.

Library of Congress Control Number: 2016935953

161718321 Printed in the United States of America

Our titles are available for adoption, license, or bulk purchase by institutions, corporations, etc. For additional information, please contact the Customer Service Dept. at 800-232-0223(toll free).

CONTENTS

INTRODUCTION

This book is written to give interested readers or first year under-graduates a comprehensive understanding of the theory of special relativity. It cannot do that using only the mathematics that first year undergraduates accumulate before their second year, and so the necessary extra mathematics is included in the book. This makes the book more than just a book on the theory of special relativity. The book is also an introduction to some of the mathematics that students will meet in the later years of their studies. The book does not give a detailed, and dry, exposition of the mathematics, but chooses only the bits of the mathematics necessary to understand the theory of special relativity. In so doing, the book gives an overview of the mathematics that students will find particularly useful when they come to study the mathematics in detail. Thus, although the book was originally intended to present only the theory of special relativity, it has turned out to also be a presentation of a little of the mathematics of physics.

The theory of special relativity is a theory of the nature of space and time and of motion through space and time. With this in mind, this book delves into the nature of empty space and the mathematics of empty space. The book introduces the reader to the geometric spaces that are derived from the mathematical objects known as the finite groups.[1] Space-time is not what mathematicians call a metric space[2], and we do not consider the metric spaces. In looking at the finite group spaces, we take the reader to the frontiers of research into the nature of space and time. We use the finite group spaces to derive the special theory of relativity from no more than the real numbers and the finite groups. We also derive Maxwell's equations of electromagnetism from the finite groups and the real numbers. At those frontiers of human knowledge, we go on to derive a 4-dimen-

---

[1.] The finite groups will be defined later.

[2.] A metric space is a mathematical function that defines the distance between two point in such a way that:

   a. The distance between points A & B is the same as the distance between B & A.

   b. The distance from A to A is zero.

   c. The distance between A & B via another point C is greater than or equal to the distance between A & B. This is known as the triangle axiom; it is the failure to satisfy this axiom that disqualifies space-time from being a metric space.

sional space-time that matches what we observe and which is more often expressed as the Lorentz group.

The first two chapters cover much of what is called special relativity without need for the mathematics. From these two chapters alone, the reader will gain a quite deep knowledge of the theory of special relativity.

The book generally follows Minkowski's geometrical approach to the theory of special relativity. Minkowski space is often absent from modern presentations of the theory of special relativity, and your author feels that those presentations are lesser presentations because of this omission. The book does, however, not rely solely on Minkowski and presents special relativity in a variety of ways. In particular, Minkowski used 4-vectors to formulate special relativity; we use 4-vectors, and we then cover the same material using matrices.

There is much repetition in the book. The same aspect of special relativity is presented in different ways and from different viewpoints. The important points are emphasised and re-emphasised. Sometimes, what has already been said once is said again when it is needed later in the book thereby ensuring that the student is ready to grasp the later material.

Of necessity in a book about special relativity, we cover the mathematics of the Lorentz transformation for which we need matrices and the hyperbolic trigonometric functions, and so we introduce the mathematics of matrices and trigonometry. The book includes the vector calculus of electromagnetism, which is normally not taught until the second year in many universities. We need the vector calculus to understand 4-vectors, which are a central part of a conventional exposition of special relativity. Having covered 4-vectors, the electromagnetic 4-potential and the electromagnetic 4-tensor follow with only a little more effort. Don't panic – all will be revealed; it is not as difficult as it sounds.

This book adopts the view of space as being derived from the finite groups rather than it being just n copies of the real numbers fixed together. To do that, we need an introduction to finite groups, which we present with matrices in a way different from, and simpler than, the usual presentation. This leads naturally to the Lorentz transform and also to quaternions. The "quaternion axiom" postulates that space-time is a quaternion structure, and there is evidence for this, and so we look at quaternions. To understand quaternions, we need to understand numbers, and so we include an understanding of division algebras and of non-commutativity. Quaternions are becoming more fashionable in physics than they once were, but, simple as they are, very few western universities teach them as standard, and so we give a brief introduction to them. We then derive the 4-dimensional structure of the space-time we see around us and electromagnetism from the finite groups.

We give a brief introduction to cosmology leading to the cosmic microwave background and the implications of this being a universal, but not absolute, reference frame.

Throughout the book, there are scattered "asides" which are comments that are not part of the main theme of the book but that link the main theme to other areas of physics and maths. Some of these asides are historical or biographical, some are simple comments, and some are of central importance to other areas of physics such as particle physics. In many ways, the asides indicate the unity of physics and mathematics. The reader need not digest the asides, and can ignore them completely without loss of understanding of the theory of special relativity, but they do add breadth to the subject.

The book contains a little of the history of the theory of special relativity and of physics and mathematics in general. This is included to lighten the load and to deepen the reader's understanding of science and of how it progresses humankind's understanding of the universe.

The book does not skimp on the mathematics, but, your author hopes, it does not bury the physics under an obscurity of technicalities. Even so, it is unlikely that a reader will digest all that is in this book in a single reading. Three readings are more normal for academic books.

As physicists and mathematicians, we quest to find a "grand unified theory of everything." We are closer to a unified field theory of the particles and forces in the universe than we have ever been, but that alone cannot be a theory of everything for it contains no understanding of the empty space and time that clearly exist in our universe. Historically, progress in this direction has been stubbornly slow, and, perhaps for lack of any idea how to proceed, mathematicians and physicists have directed their efforts, very profitably and very understandably, elsewhere. Recent developments lead us to think that it is possible that humankind will come to understand space and time in the near future, and we hope to take a step towards that understanding of space and time with this book.

Your author hopes he has produced an enjoyable book. He hopes the book will be easy to read and deeply interesting to the curious reader. Your author also hopes that this book will engender within the reader a life-time interest in the nature of the empty space that surrounds us and its relationship to the particles and forces of the universe.

Dennis Morris
May 2016

1

AN OVERVIEW OF THE THEORY OF SPECIAL RELATIVITY

1.1 PHYSICS IS INVARIANT UNDER ROTATION

If we take a kettle full of water and point the kettle's spout westward, the water in the kettle boils at 100° centigrade. If we then turn the kettle to point its spout northward, the water in the kettle still boils at 100° centigrade. The temperature at which a kettle boils water does not change with the spatial direction in which its spout points. (We ignore extraneous effects such as air pressure.) It would astound us if physical effects did differ with change in spatial direction of the physical system. Imagine how weird a car would seem if its engine worked only when the car pointed northward or how weird sugar would be if it sweetened tea only if the teacup handle was pointed westward. We believe, both from observation and for good theoretical reasons, that the physics of the universe is independent of direction in space. This is not only true upon the surface of the Earth; this is true, we believe, everywhere in the universe. We say that, "the boiling point of water is invariant under rotation in space". The universe is the same in all directions – it is isotropic in space. We say that physics is invariant under rotation in space. We mean that the way things work - the physical laws of the universe - do not change when we alter the direction in space of the physical system. This (blindingly obvious) understanding is central to the special theory of relativity. Indeed, apart from the details, this understanding is the special theory of relativity. The remainder of this book is just the details. We repeat and embolden:

The physics of the universe is invariant under rotation.

1.2 SPACE AND TIME ARE NOT SEPARATE THINGS

To Isaac Newton (1642–1727), as to all humankind except modern theoretical physicists and modern philosophers, time was a thing separate from space. Time is different from space. Time never stops flowing, and we never stop moving through time, but it is easy for us to stop moving through space. Newton saw time as a single entity complete on its own and entirely separate from space. Newton saw space as a thing in its own right that was separate from time. Newton lived a good while ago. Since then, due to Albert Einstein (1879–1955) and others, our understanding of time and space has changed. We now see time as a dimension in space-time; we see space as a dimension (or three[1]) in space-time. We are of the view that space and time are a single entity. Space and time are as Romeo and Juliet in Shakespeare's play "Romeo and Juliet". Without time, there would be no space-time, and, without space, there would be no space-time. Without Romeo, there would be no Romeo and Juliet, and, without Juliet, there would be no Romeo and Juliet. Space and time are not separate entities but are tied together into a single (2-dimensional to start with) space-time. Perhaps this statement should be repeated and emboldened:

Space and time are not separate entities but are joined together into a single space-time.

---

[1] For the first part of this book, we will treat space-time as being 2-dimensional. We will adapt to it being 4-dimensional in the latter part of the book.

1.3   MASS DIMENSIONS

If space-time is one entity, then time must be the same stuff as space; even though the two things appear to be different to we humans. We say that the mass-dimensions of space and time are the same. We write this as:

$$[T] = [L] \tag{1.1}$$

What we are saying here is that, if space is measured in meters, then time must be measured in meters. Alternatively, if time is measured in seconds, then space must be measured in seconds (think light-second or light year). Instead of saying that the sun is 93 million miles from Earth, we say the sun is 8 seconds from Earth. A consequence of this is that the mass-dimension of velocity is just a number:

$$[Velocity] = \frac{[Space]}{[Time]} = \frac{[Meters]}{[Meters]} = \frac{[Secs]}{[Secs]} = 1 \tag{1.2}$$

It is normal in theoretical physics to set the velocity of light, which, being a velocity, is just a number, equal to unity (one). They do this because it is easier to put $c = 1$ in formulae and then not worry about factors of c. The factors of c are easily recovered when needed. Such units with $c = 1$ are sometimes called geometrical units by relativists.

The concept of mass-dimension appears almost everywhere in theoretical physics. The equivalence of space and time is the basis of this concept of mass-dimension.

1.4   DIRECTIONS IN SPACE-TIME

Different directions in space-time are different velocities. Since space-time is a single (2-dimensional for us to start with) entity, it has one space dimension and one time dimension. A direction in any type of space is a ratio of one dimension to another dimension. (Think $gradient = \frac{y}{x}$ on a sheet of graph-paper.) A ratio of space to time is a velocity, like meters per second. In space-time, different slopes (gradients) correspond to different directions which correspond to different velocities.

The reader might like to picture a 2-dimensional flat sheet of paper with two axes drawn upon it.

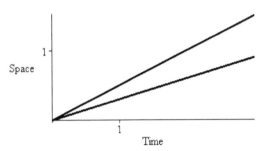

One axis is the time axis and the other axis is the space axis. This is space-time. The slope (gradient) of the lines from the origin is given by the ratio of the two axes, $\dfrac{Space}{Time}$, but this is velocity. Different slopes are different directions, both on the sheet of paper, and in space-time. These different directions are different ratios of the space dimension and the time dimension. A ratio of space to time is a velocity, and so the different directions in space-time are just different velocities. We see that what appears to us to be different velocities are no more than different directions (like East or South-east) in space-time. Perhaps this should be repeated and emboldened.

Different velocities are no more than different
directions in space-time.

If we take a kettle full of water that is stationary before us, it boils at 100° centigrade[2]. If we then put the kettle on to a train traveling at 100,000 kilometers per second, the kettle still boils at 100° centigrade. The temperature at which a kettle boils water does not vary with the velocity at which the kettle moves. A particular velocity is just a particular direction in space-time, and so, a kettle moving at 100,000 kilometers per second is just a kettle with its spout pointing in a

---

[2]. That the boiling point of water varies with height above sea level or other effects is irrelevant to this discussion.

different direction in space-time to the direction pointed by the spout of a kettle that is stationary. We say that, "the boiling point of water is invariant under rotation in space-time", this is just a way of saying that the physics of the universe is the same at all velocities – space-time is isotropic. We will repeat and embolden this statement also.

> *The physics of the universe is invariant under rotation in space-time.*
> *The physics of the universe is the same at all velocities.*

We mean that the way things work - the physical laws of the universe - do not change when we alter the velocity at which a physical system travels. As your author writes, the Voyager space-craft, launched by NASA in the 1970s, is moving away from Earth at approximately 61,400 km/hr and all the electronic systems within it are working exactly as they did when it was stationary here on Earth in the early 1970's. Astronomers can detect stars moving very rapidly away from the Earth. These stars still shine in the same way that our own sun shines. Everything works the same regardless of the velocity at which a physical system moves. This is the special theory of relativity, and that deserves emboldening.

> *Everything works the same regardless of the velocity*
> *at which a system moves.*

In essence, the special theory of relativity is no more than a statement that space-time is a single entity (space and time are not two things) and that space-time is isotropic.

The Special Theory of Relativity

Space-time is a single entity. Space and time are not separate.
Space-time is isotropic

Kettles boil at the same temperature regardless of the direction in which they are pointed (in space-time). Kettles boil at the same temperature regardless of the velocity at which they are moving.

1.5 THE CONSTANCY OF THE SPEED OF LIGHT

The temperature at which water boils depends upon the strength of the electromagnetic forces that hold the water molecules close to each other. The strength of the electromagnetic force depends upon the electrical permittivity, ε_0, and the magnetic permeability, μ_0, of empty space. The speed of light also depends upon these two physical constants; it is given by:

$$\text{speed of light} = c = \frac{1}{\sqrt{\varepsilon_0 \mu_0}} \tag{1.3}$$

Thus, the temperature at which water boils depends upon the speed of light; or the speed of light depends upon the temperature at which water boils; or they both depend upon how easily electric fields and magnetic fields can penetrate empty space, or they are all just a different view of the same thing. Thus, since kettles boil at the same temperature at all velocities, the speed of light is the same at all velocities. That's simple enough! The speed of light is just a law of physics, and so it should be the same at all velocities.

Aside: μ_0 is also called the vacuum permeability, the permeability of free space, or the magnetic constant. It is derived from Ampere's law, and its value was fixed in 1948 by the definition of the ampere. That value is $\mu_0 \approx 1.257 \times 10^{-6} \ Hm^{-1}$ (Henrys per meter) in SI units. μ_0 is a measure of how easily magnetic fields can penetrate empty space.

ε_0 relates units of electrical charge to mechanical quantities. It is defined to be $\varepsilon_0 = \dfrac{1}{c^2 \mu_0} \approx 8.85 \times 10^{-12} \ Fm^{-1}$ (Farads per meter) in SI units. ε_0 is a measure of how easily electrical fields can penetrate empty space. Each type of substance has its own measure of how easily electrical fields can penetrate it; this measure is called the relative permittivity. For olive oil, the relative electrical permittivity is $\varepsilon_r \approx 3 \ Fm^{-1}$. For ice ($-2°C$), the relative electrical permittivity is $\varepsilon_r \approx 94 \ Fm^{-1}$.

Now imagine a physics student on a railway platform making a cup of tea and shining a torch along the railway line towards an oncoming train that is moving towards the platform at 100,000 kilometers per second. The physics student on the platform watches the water for his tea boil at 100° centigrade and measures the velocity of the light leaving his torch to be 300,000 kilometers per second away from him. On the train, there is a mathematics student who is also making a cup of tea. The mathematics student watches the water for her tea boil at 100° centigrade and measures the light from the physics student's torch to be moving at 300,000 kilometers per second towards her. (If her kettle boils at 100o centigrade, light must move at 300,000 kilometers per second. We've just worked that out.) What happened to the 100,000 kilometer per second velocity of the train? Ought not the mathematics student to measure the light from the physics student's torch to move at the's velocity of 400,000 kilometers per second? No, she ought not to because, if she did, her kettle would have to boil at 175° centigrade to match the change in the values of the electrical permittivity, ε_0, and the magnetic permeability, μ_0.

An old Sioux Indian once told your author that this "loss of 100,000 kilometers per second" seems counter-intuitive. Your author is of the opinion that it actually is counter-intuitive. None-the-less, it is true. In 1968, Farley, Bailey, & Picasso measured the speed of radiation emitted when π-mesons decay. Although the mesons were moving at close to the speed of light through the laboratory, the light emitted by them was measured to be c in the laboratory[3]. Both the mathematics student on the train and the physics student on the platform measure the beam of light to be moving at the same velocity because the laws of physics (of which the speed of light is one) are the same in all directions in space-time – that is the laws of physics are the same at all velocities.

Let us suppose that the mathematics student on the train did measure the speed of light from the physics student's torch to be 400,000 kilometers per second, and, at the same time, she also measured the speed of light traveling across the carriage, from a lamp in the carriage, to be 300,000 kilometers per second. Would not this mean that the kettle boils at different temperatures depending upon whether its spout points across the carriage or its spout points along the carriage? Suppose the train goes around a bend in the track. Suppose the mathematics student measures, from the lamp in the carriage, the speed of light in the direction of her travel and gets 300,000 kilometers per second; together with the 400,000 kilome-

---

[3.] Farley, Bailey, & Picasso in Nature, 217, 17 (1968).

ters per second result from the light of the physics student's torch, there would be two different values for the speed of light in the same direction. The poor kettle would not know which way to turn.

Imagine two observers both measuring the speed of light in both northward and westward directions. Spatial isotropy means that the speed of light in both directions is the same. Now suppose the observers are moving relative to each other in the northward direction but are stationary relative to each other in the westward direction. Because they are stationary relative to each other in the westward direction, they will agree on the speed of the westward moving light. Since they agree on the speed of the westward moving light, they must agree (they each have spatial isotropy) on the speed of the northward moving light – if they do not so agree, then we have lost spatial isotropy. We see that invariance of physics under rotation in space necessitates invariance of physics under rotation in space-time. If light is to move at the same speed northward as it does westward for all observers, then that speed must be independent of the relative northward velocities of those observers.

Many authors of books on the theory of special relativity start from the constancy of the speed of light under change of velocity. Indeed, this is part of what led Einstein to special relativity (the other part is magnetic fields caused by moving electric charges). From the constancy of the speed of light, those authors deduce many of the counter-intuitive aspects of special relativity. We will follow their reasoning later in this book. However, in spite of the constancy of the speed of light being the starting point for many authors, it is not basic to the theory of special relativity. In fact, as we will see later, the speed of light is not unique to light; we will see later that all things, including we humans, travel through space-time at the speed of light. The invariance of the speed of light with velocity is no more than the general invariance of physical phenomena under rotation in space-time.

Your author has rattled on about physics being invariant under rotation in space-time (and space-space), but he has said nothing about invariance under translation in space and in time. We believe that the physics of the universe is the same ten billion light-years from Earth as is here on Earth; we believe that kettles boil at 100°C in distant galaxies just as they do on Earth. This is the invariance of physics under translation in space. We also believe that the physics of the universe was the same ten billion years ago as it is now and that it will be the same ten billion years into the future as it is now; we believe that kettles will boil at 100°C in ten billion years time just as they do today. This does not mean that the universe was the same ten billion years ago as it is now; it means only that the universe works in the same way now as it did ten billion years ago. This is invariance under translation in time. Such translational invariance it is a belief and not a fact; it might be that the speed of light was infinite at the start of the universe and has lessened since then; only observation can decide, and no-one has been ten billion years into the future or ten billion light-years into space.

Your author has written of the unity of space and time. Later in this book, we will see that within special relativity we also have the unity of many other concepts such as momentum and energy, force and power, and electric fields and magnetic fields. However, your author must now point out that this unity breaks for the stationary observer. For the stationary observer, space and time are separate things, as are momentum and energy, and force and power, and electricity and magnetism. It is only when we consider systems in motion relative to ourselves that we see the unity of these things. We need to accept the unification of these things when we deal with the physics of systems that are in motion relative to ourselves. We do not need to accept this unification when dealing with systems that are stationary relative to ourselves. Think about it; is it not obvious to you, as you sit there, that space is separate from time as far as you are concerned?

Having said all the above about time being a dimension in space-time, let us remember that time is not space. We can, and will, treat time mathematically as if it is were a dimension in space. This works, but it does not mean that time is space. Every color in the world is a mixture of three primary colors, red, blue, and green. I can thus specify any color by just three ordered numbers; these numbers being the proportions of each of the primary colours in the color that I am specifying. Similarly, I can specify any position in a 3-dimensional space by three ordered numbers (a vector), but, even though color is mathematically identical to 3-dimensional space, this does not mean that color is a 3-dimensional geometric space. The same is true about including time as one number in a set of four numbers that specifies a position in the universe. Just because the math works like a 4-dimensional space does not mean that it is a 4-dimensional space. We will see later that the trigonometric functions of 2-dimensional Euclidean space are the cos( ) and the sin( ) functions and that other than a displacement of 90°, these two functions are identical. We will see later that the trigonometric functions of 2-dimensional space-time are the cosh( ) and the sinh( ) functions. These functions are very different from each other; this is why time is different from space.

We've been a little repetitive in this chapter. We have repeated the same concepts several times. It is partly because of the importance of understanding these concepts that we have been so repetitive. It is partly because of the difficulty of understanding the consequences of these concepts that we have been so repetitive. Within this chapter, there are all the difficult basic concepts of the theory of special relativity. There are other concepts in special relativity that are difficult, but they are not basic concepts. We repeat:

> *Space and time are just two different dimensions in one type of space called space-time.*
>
> *Different velocities are no more than different directions in space-time.*
>
> *The physics of the universe is invariant under rotation (change of direction) in space-time (and space-space).*

Aside: To be acceptable, any theory of particle physics must be invariant under "Lorentz transformations". Lorentz transformations are just rotations in space-time, and so, to be acceptable, any theory of particle physics must be invariant under rotation in space-time (the same in all directions in space-time – the same at all velocities). We knew that.

However, the particle physics theory must also be invariant under three other types of transformations known as $\{U^{(1)}, SU^{(2)}, SU^{(3)}\}$. These transformations are, respectively, rotation in $\mathbb{C}^1$ space (the complex plane), rotation in $\mathbb{C}^2$ spa (two copies of the complex plane fitted together like $\mathbb{R}^2$), and rotation in $\mathbb{C}^3$ space (three copies of the complex plane fitted together like $\mathbb{R}^3$).

1.6   SYMMETRY AND NOETHER'S THEOREM

In the first paragraph of this chapter, we assumed that a kettle boils at the same temperature regardless of the direction in which its spout is pointing. The reader did not question this "blindingly obvious" fact, but why is it true? In the first half of the 20th century, Emmy Noether, an outstanding mathematician, was able to prove the truth of this mathematically. In doing so, she was also able to show why some things (energy, angular momentum,...) are conserved quantities in physics. The mathematics of what she did is beyond the remit of this book, but we will briefly overview what she achieved.

Physicists define a symmetry as being a change in perspective that leaves the equations of physics unchanged. Rotation in space-time is such a change in perspective, as is rotation in space – thus rotation in space-time is a symmetry. Another example is translation in space which just moves the origin of the co-ordinate system. Yet another is translation in time. Mathematically, a symmetry is a variation to the fields in the Lagrangian (the Lagrangian is a mathematical expression) that leaves the equations of motion invariant. This is a precise, but difficult, way of saying the laws of a physical system do not change under a symmetry. The form of the Lagrangian stays the same under a symmetry (a rotation or a translation, say).

Aside: Lagrangian mechanics is a reformulation of classical mechanics based upon the concept of a stationary action, A. This reformulation was introduced by Joseph-Louis Lagrange (1736–1813) in 1788. It applies to systems whether or not they conserve energy and momentum and it gives the conditions under which energy and momentum are conserved in those systems. The Lagrangian, L, is the difference between the total kinetic energy and the total potential energy:

$$L = T - V \qquad (1.4)$$

Where T is the total kinetic energy and V is the total potential energy. From the Lagrangian, we can calculate the equations of motion of the system that describe how the system will evolve through time. The path integral of the Lagrangian, L, is called the action, A:

$$A = \int dt\, L \qquad (1.5)$$

Symmetry transformations (for example, rotations) leave *A*
unchanged (invariant). Such rotations form a group of infinite order
that is called a Lie group.

Symmetries are associated with conservation laws. The conservation
of momentum is associated with physics being unchanged by a translation in space. The conservation of momentum is a statement that space
is homogeneous. The conservation of energy is associated with physics
being unchanged by a translation in time – kettles boiled at 100° centigrade for Euclid in 300BC. The conservation of energy is a statement
that time is homogeneous. The conservation of angular momentum is
associated with physics being unchanged by a rotation in space. This
states that space is isotropic. All of the above was proven mathematically
by Emmy Noether and is expressed in Noether's theorem.

_____

NOTE

*In the paragraph above, we have said that conserva-
tion of momentum is associated with translation in
space and that conservation of energy is associated
with translation in time as if space and time were separate things. We will see later that, in special relativity, we actually have the single law of conservation of
momenergy. Momenergy is momentum and energy.
This is associated with translation in space-time. In
the theory of special relativity, the two conservation
laws are united into one conservation law as space
and time are united into one space-time.*

1.7   NOETHER'S THEOREM[4]

For every continuous symmetry of the Lagrangian, there is a
conserved current given by:

$$J^{\mu} = \frac{\partial L}{\partial\left[\partial_{\mu}\varphi\right]}\delta\varphi \qquad (1.6)$$

_____

[4.] This theorem was published by Noether in *Invariante Variations probleme in
1918 and was presented by Felix Klein to the Royal Society of Gottingen. Noether
could not present it herself because women were not allowed membership of that
society in those days.*

And there is a conserved charge associated with the conserved current given by:

$$Q = \int d^3x \, J^0 \qquad (1.7)$$

You do not need to understand this theorem to understand the theory of special relativity, but you will meet it later in your studies. It is one of the most important results of theoretical physics. It can be seen as the mathematical proof that a kettle boils at the same temperature regardless of the direction of its spout.

Aside: Emmy Noether was born into a family of eminent mathematicians on March 23rd 1882 in Erlangen, Bavaria. She finished school at the age of 15, but was not allowed to attend university because she was a woman – yes humankind really was that backward only a few decades ago! However, her father, Max Noether, was the professor of mathematics at the University of Erlangen, and, it seems, together with, Paul Gordan, that he allowed Emmy to attend the lectures there. In 1903, exceptionally, and with the support of the mathematicians there, she was allowed to attend lectures at Gottingen given by Minkowski, Klein, and Hilbert, and, in 1904, very exceptionally, she was allowed to enrol as a student at Gottingen university. It is a matter of which mathematicians and physicists may be proud that through history many of their number have strongly opposed discrimination based on both gender and race. As long as the mathematics is correct, the mathematician can hail from the Andromeda galaxy, be bright green with yellow dots, have six heads, and be of three different genders at the same time as far as we are concerned·

In 1908, Emmy produced a doctoral thesis on invariance[5], following which she continued to do research at Erlangen on a more abstract approach to the theory of invariants but, being a woman, was unpaid for her work. She was eventually invited to join Hilbert at Gottingen, but the university refused to allow her to teach because she was a woman. It was this that led to Hilbert's famous, "..this is after all an academic institution not a bath-house" outburst. Hilbert

---

[5.] On Complete Systems of Invariants for Ternary Biquadratic Forms.

got around the problem from 1916 to 1919 by letting her deliver lectures in his name advertised as being delivered by him, but she remained unpaid. From 1919 to 1923, she was allowed to lecture in her own name, but was unpaid, and, in 1923, she was finally allowed to take a bona fide, but still unpaid, university position. She was dismissed from this position in 1933 by the Nazis because she was Jewish. She fled to the USA where she died from infection following minor surgery in 1935. The theorem above, which is widely recognised as possibly the most important theorem in theoretical physics, carries her name.

WORKED EXAMPLES

1. Taking time to be of the same mass-dimensions as space, what are the mass-dimensions of velocity?

    Ans: We have: $v = \dfrac{L}{T}$. With $[L] = [T]$, this gives $[v] = \mathbb{R}$, which means that the mass-dimension of velocity (like the velocity of light for example) is just a number. Velocity is without mass-dimensions.

2. Taking time to be of the same mass-dimensions as space, what are the mass-dimensions of energy?

    Ans: We have: $E = \dfrac{1}{2} mv^2$. With $[L] = [T]$, this gives $[E] = [\mathbb{R}M\mathbb{R}^2] = [M]$, which means that the mass-dimension of energy is mass.

EXERCISES

1. What are the mass-dimensions of acceleration?
2. What are the mass-dimensions of force?

THE RESULTS OF SPECIAL RELATIVITY WITHOUT DETAILED EXPLANATION

The fact that all laws of physics are the same at all velocities has many consequences. We will study the details later. For now, we will merely list the more important consequences. The reader should note that an observer does not have to be a human being but could be an inanimate object such as a photographic plate or an electron. We point out that a stationary observer is stationary – they are not moving – they are stood still. A moving observer is moving relative to a stationary observer.

Special relativity is essentially a 2-dimensional theory. Many text books dress it up as being 4-dimensional so that it fits into the 4-dimensional space-time that we observe around us and because this is necessary to fit with the electromagnetic field tensor. Such dressing up is done by adding two inert spatial dimensions into the mathematics and then taking them along for the ride. The reader will lose nothing if she thinks of special relativity in two dimensions only. The 4-dimensional space we observe is described by a different theory known as the Lorentz group, which is compatible with special relativity. We consider the Lorentz group towards the end of this book.

1. *Velocity is Relative:* **All observers can, and do, consider
    themselves to be stationary.** A math student on a railway plat-
    form watching a physics student on a train passing through
    the station can take the view that she, the math student, is
    stationary and that the physics student on the train is moving.
    A physics student on a train that is passing through a station
    can take the view that he, the physics student on the train, is
    stationary and that the math student standing on the platform
    is moving. Both views are correct physics. The difference is
    simply the arbitrary way (direction) in which the different
    observers align their space-time axes.

    The time axes of the two students are not aligned. The origins
    coincide, but their axes are at an angle (a space-time angle)
    to each other. The size of the angle is the magnitude of their
    relative velocity, and the greater the space-time angle between
    their time axes, the greater the velocity difference between the
    two students – space-time angle is relative velocity. The phys-
    ics student thinks he has aligned his time axis "horizontally"
    and that the time axis of the mathematics student is at an angle
    to his "horizontal". The mathematics student thinks she has
    aligned her time axis "horizontally" and that the time axis of
    the physics student is at an angle to her "horizontal".

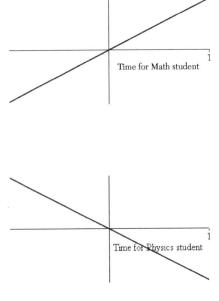

Time for Math student

Time for Physics student

Relative velocity is no more than this non-alignment of space-time axes.

2. *The Relative Velocity Difference Between Observers is Agreed by the Observers:* If a math student on a railway platform measures the velocity of a physics student on a train to be v relative to herself, then the physics student on a train will measure the velocity of the math student standing on the platform to also be v relative to himself. They both agree that the velocity at which they pass each other is v. That is to say that they both agree on the magnitude of the space-time angle between their time axes.

3. A. *Time Dilation:* The rate at which things happen in moving spaceships (say the heart-beats of the spaceship's crew) appears to the stationary observer to lessen (the heart-beats of the spaceship's crew appear to the stationary observer to slow down). What is really happening is that the time between heart-beats is being stretched. We call this time dilation. We have:

$$\Delta t' = \Delta t_0 \frac{1}{\sqrt{1 - \frac{v^2}{c^2}}} = \Delta t_0 \gamma \tag{2.1}$$

Where Δt_0 is the length of time taken for a process (say a stationary person's heart-beat) in the "stationary world" as seen by the stationary observer and $\Delta t_{\prime}$ is the corresponding length of time taken for the same process (a moving person's heart-beat) in the "moving world" also as seen by the stationary observer. The relative velocity of the two observers is v, and c is the speed of light. The expression with the square root is known as gamma:

$$\gamma = \frac{1}{\sqrt{1 - \frac{v^2}{c^2}}} \tag{2.2}$$

When $v = 0.9c$ and $.\Delta t_0 = 1.$, we have:

$$\Delta t' = \frac{1}{\sqrt{1 - \frac{(0.9)^2 c^2}{c^2}}} = \frac{1}{\sqrt{1 - 0.81}} = \frac{1}{0.436} = 2.29 \tag{2.3}$$

So, when the stationary observer sees a heart-beat take one second in the "stationary world" according to the stationary clock, she sees a heart-beat in the "moving world traveling at $v = 0.9c$" take 2.29 seconds according to the stationary clock. We emphasize "according to the stationary clock" in the previous sentence. As judged by stationary observers, processes proceed more slowly when they proceed in moving spaceships (or trains, or ships, or cars, or planes, or...). To an observer moving with the spaceship, everything seems normal aboard the spaceship.

The mass media often present this phenomenon as "Time slows down in moving spaceships". It is the processes of the universe that slow down in moving spaceships, and so it takes more "stationary time" from the start of the moving process to the end of that moving process as seen by a stationary observer. Thus, it seems to a stationary observer that the "moving world" slows down. A part of this "moving world" is the mechanism of the moving clock, and so the moving clock appears, to the stationary observer, to slow down; there is more stationary time between the ticks of a moving clock than between the ticks of a stationary clock. This is more than only appearance; the moving clock really does slow down compared to the stationary clock. We emphasize that what is really happening is that the time between the ticks of a moving clock is stretched (dilated). We repeat, to an observer moving with the spaceship, everything seems normal aboard the spaceship.

For all practical purposes, all we need to know is that the processes of the universe slow down in moving reference frames.

Since two observers moving relative to each other can both consider themselves to be stationary, time dilation seems contradictory since both observer's clocks will each run slower than the other. Each observer has their own clock and each thinks that the other's clock is running slowly. Each is correct. Think of it as a female observer having her (right-angled) co-ordinate axes horizontal and vertical while a male observer has his (right-angled) axes at 45° to the horizontal and at 45° to the vertical.

Now imagine a meter stick along the (female) horizontal axis. The female observer will say that the stick is one meter long in the x-direction while the male observer will say the stick is only $\frac{1}{\sqrt{2}}$ meter long in the (his) x-direction. If the male observer lays a meter stick along his x-axis (at 45° to the horizontal), he will say the stick is one meter long in the x-direction but the female observer will say that his stick is only $\frac{1}{\sqrt{2}}$ meter long in her x-direction. They are both correct, and each stick is shorter than the other.

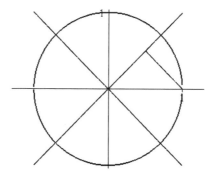

Only if one of the observers adjusts their co-ordinate system (rotates it by 45° to match the co-ordinate system of the other observer) will they be able to agree on the lengths of the sticks – but there is no contradiction. In space-time, such a rotation is a change of velocity. The observers will agree only if one changes their velocity to be the same as the velocity of the other. Any necessary adjustments are made when one or both of the observers alter their velocity to match the velocity of the other – when one or both rotate in space-time until they both point in the same direction in space-time. The seeming contradiction is a consequence of using differently orientated space-time co-ordinate systems. There will be more on this later; it is the source of a lot of seeming paradoxes in special relativity theory.

B. *Clocks slow down when they are in gravitational fields:* This is a result of general relativity and nothing to do with special relativity. Clocks near to the center of the Earth (stronger gravity) run slower than clocks that are far from the center of the Earth[1] (weaker gravity)[2].

C. *Non-gravitational acceleration does not cause time dilation:* Thinking of velocity as rotation in space-time, we would expect acceleration by itself not to cause time dilation. This is verified by experiments that have shown acceleration does not have any time dilation effects other than the time dilation that results from relative velocity. CERN have shown that an acceleration of $10^{18}g$ experienced by muons circulating in a storage ring does not add to the time dilation that muons experience due to their relative velocity. Vessot[3] *el al* used hydrogen maser clocks in rockets to determine the velocity dependent time dilation of special relativity and the time dilation of a gravity field and any time dilation due to non-gravitational acceleration to an accuracy of 10^{-4}. They found no time dilation due to the non-gravitational acceleration of the rocket. Thus, gravity does cause time dilation but non-gravitational acceleration does not.

4. *Length Contraction:* The length of a moving object appears to the stationary observer to be less than the length of the same object when it is stationary. We call this length contraction. We have:

$$\Delta l' = \Delta l_0 \sqrt{1 - \frac{v^2}{c^2}} = \frac{\Delta l_0}{\gamma} \tag{2.4}$$

Where Δl_0 is the length in the "stationary world" as seen by the stationary observer and $\Delta l'$ is the corresponding length in the "moving world" also as seen by the stationary observer.

---

[1] This is not quite true. The sea level at the poles is closer to the Earth's center that the sea level at the equator, but clocks run at the same rate at sea level throughout the world. This is because sea level is a gravitational equipotential surface.
[2] If this effect was ignored, the GPS system would not work.
[3] R.F.C. Vessot *et al* Phys. Rev. Lett. 45, 2081 (1980).

The relative velocity of the two observers is v, and c is the speed of light. When $v = 0.9c$ and $\Delta l_0 = 1$, we have:

$$\Delta l' = \sqrt{1 - 0.81} = 0.436 \qquad (2.5)$$

So, what the stationary observer sees as a rod of length one meter in the "stationary world', he sees as a rod of length 0.436 meters in the "moving world traveling at $v = 0.9c$'. Imagine that light travels at one meter per second; it is only a matter of which units humankind chooses to use. If, in the stationary observer's view, the process of light traveling from one end of a meter rod to the other end of the meter rod takes one second in the "stationary world", then, because of the time dilation effect, the same process will take 2.29 seconds in the "moving at $v = 0.9c$ world', as seen by the stationary observer, and so the light will travel only $\dfrac{1}{2.29} = 0.436$ moving meter in one stationary second – that is 0.436 moving meters per stationary second. But light always travels at one meter per second. Therefore, the 0.436 meter length must correspond to one meter of stationary length, and so, length in the "moving world" must appear to the stationary observer to be less than it is in the "stationary world" (0.436 meters to one meter at $v = 0.9c$). Time dilation plus constancy of the speed of light equals length contraction. This length contraction is also known as Lorentz-Fitzgerald contraction or Lorentz contraction.

Length contraction is most evident in the length contraction of a wire carrying an electrical current (moving electrons). The distance between the electrons contracts and thus increases the electron density in the wire and thus the electromagnetic force from the wire. This increased bit of the electromagnetic force is the magnetic force that emanates from an electrical current. Thus, without length contraction, we would have no magnetic fields and so no electric power generators.

5. *Limiting Velocity:* There is a finite velocity (it is the velocity of light) that is the upper bound of the velocity at which any object can move through space (not space-time). Nothing

can move faster through space than this greatest velocity within space-time. Since increase in velocity is just rotation in space-time, this means there is a limiting asymptote towards which we rotate and past which we cannot rotate. This is very different from the more familiar rotation in the Euclidean plane which is completely unfettered.

Without a limiting velocity, objects could move infinitely quickly from place to place and thereby be in two, or many, places at the same time. This is called non-locality and it does seem to occur in quantum theory regarding the spin of photons. General non-locality of everything would produce a radically different universe from the one we observe. Hence, it is commonly thought to be a good thing that we have a limiting velocity.

6. *Addition of Velocities is Counter-Intuitive:* If a math student is moving relative to a stationary biology student at velocity u_1, and a physics student is moving in the same direction as and relative to the math student at velocity, u_2, then the stationary biology student will see the velocity of the physics student to be less than the sum $u_1 + u_2$. We have:

$$v' = \frac{u_1 + u_2}{1 + \frac{u_1 u_2}{c^2}} \tag{2.6}$$

Where v' is the velocity at which the physics student moves relative to the stationary biology student. Since we have a limiting velocity, we must not be able to simply add velocities for this would enable us to exceed the limiting velocity.

7. *Acceleration is Non-Newtonian:* In Newtonian mechanics (the mechanics we are used to and the mechanics that engineers use), if a rocket engine causes a rocket to accelerate at a particular rate when the rocket is moving slowly, it will drive acceleration of the rocket at the same rate when the rocket is moving rapidly. This invariance of acceleration with velocity, together with the invariance of force with velocity, and the invariance of mass with velocity are basic to Newtonian mechanics. In special relativity mechanics, this is not so; acceleration varies with velocity. Because there is a limiting

velocity in the universe, acceleration has to "weaken" as the velocity approaches that limiting value. There are two different cases.

For acceleration parallel to velocity, we have:

$$a' = \frac{1}{\gamma^3} a_0 = \left(\sqrt{1 - \frac{v^2}{c^2}}\right)^3 a_0 \tag{2.7}$$

Where v' is the acceleration in the "stationary world" as seen by the stationary observer and $a¢$ is the corresponding acceleration in the "moving world" also as seen by the stationary observer. a_0 is the acceleration that a person on a moving spaceship would feel. a' is the acceleration that a stationary observer would see the moving observer feel. The relative velocity of the two observers is v, and c is the speed of light. When $v = 0.9c$, we have the case:

  i) acceleration parallel to velocity:

$$a' = \frac{1}{\gamma^3} a_0 = \left(\sqrt{1 - \frac{v^2}{c^2}}\right)^3 a_0 = (0.08)a_0 \tag{2.8}$$

For acceleration perpendicular to velocity, we have:

$$a' = \frac{1}{\gamma^2} a_0 = \left(\sqrt{1 - \frac{v^2}{c^2}}\right)^2 a_0 \tag{2.9}$$

There is a γ difference. When $v = 0.9c$, we have the case:

ii) acceleration perpendicular to velocity:

$$a' = \frac{1}{\gamma^2} a_0 = \left(\sqrt{1 - \frac{v^2}{c^2}}\right)^2 a_0 = (0.2)a_0 \tag{2.10}$$

8. *Newtonian Vectors are replaced by 4-vectors:* In special relativity, space and time are stuck together and so we use a type of vector that incorporates both space and time components. Such vectors are called 4-vectors. In doing this, we are taking a "God's eye view" of the universe rather than a stationary observer's point of view. There is no accepted notational convention for 4-vectors; we choose to write them with two lines

atop a capital letter. 4-vectors have four components and a 4-vector inner product (the dot product) calculated as:

$$\left\langle \overline{\overline{X}}_1, \overline{\overline{X}}_2 \right\rangle = \overline{\overline{X}}_1 \bullet \overline{\overline{X}}_2 = t_1 t_2 - x_1 x_2 - y_1 y_2 - z_1 z_2 \qquad (2.11)$$

Notice the minus signs.

Almost all the laws of physics are expressible using scalars (real-numbers), 4-vectors, and 4-tensors. The only exceptions are in particle physics where we use spinors as well, but that is not our concern in this book.[4]

Aside: The Dirac equation that describes a relativistic electron with spin is Lorentz invariant but is expressed using spinors not 4-tensors or 4-vectors or scalars. Spinors are needed because the wave function is written with complex numbers.

Spinors are objects that define a $\mathbb{C}^2$ (two complex dimensions) sub-space of the $\mathbb{C}^3$ space (three complex dimensions)[5].

---

9. *We do not Differentiate with Respect to Time:* In Newtonian mechanics, we differentiate distance with respect to time to get velocity and acceleration:

$$v = \frac{ds}{dt} \qquad a = \frac{d^2 s}{dt^2} \qquad (2.12)$$

Since time varies (dilates) from one velocity to another, it would introduce complications to differentiate with respect to time. Instead, in the standard presentation of special relativity mechanics, we differentiate with respect to the invariant interval, τ. This is for convenience rather than for some great mathematical reason. The invariant interval is the square root of the norm of the algebra (the length of a vector). With $\overline{\overline{R}}$ representing the displacement 4-vector, this gives the 4-velocity as:

$$\overline{\overline{U}} = \frac{d\overline{\overline{R}}}{d\tau} \qquad (2.13)$$

[4.] See: Dennis Morris - The Naked Spinor ISBN: 978-1-507817995
[5.] Elie Cartan. The Theory of Spinors. 1966 – First published as "Lecons sur la theorie des spineurs'. 1937

and the 4-acceleration as:

$$\overline{\overline{A}} = \frac{d\overline{\overline{U}}}{d\tau} \qquad (2.14)$$

Since the length of a vector is unchanged by rotation of that vector, the invariant interval is invariant under rotation in space-time (change of velocity), which is why it is called invariant. In two different reference frames, which are just different rotational orientations, the invariant interval is the same, and we have:

$$\tau^2 = t^2 - x^2 - y^2 - z^2 = t'^2 - x'^2 - y'^2 - z'^2 = \tau'^2 \qquad (2.15)$$

At zero velocity, $\tau = t$, and we get the Newtonian mechanics.

10. *Force is Non-Newtonian:* Force, like acceleration, also varies with velocity. In Newtonian mechanics, force is defined in two different but equivalent ways. One way is as mass multiplied by acceleration – $F = ma$; the other way is as the rate of change of momentum with respect to time - $F = \frac{dp}{dt}$. In special relativity, force is defined in only one way. That way is based upon the rate of change of momentum, but we use 4-vectors and we differentiate with respect to the invariant interval, τ, rather than with respect to time:

$$\overline{\overline{F}} = \frac{d\overline{\overline{P}}}{d\tau} \qquad (1.16)$$

It might be better to define force as rate of change of energy with respect to space-time so that the force 4-vector would be:

$$\overline{\overline{F}} = \left[\frac{dE}{dx} \quad \frac{dE}{dy} \quad \frac{dE}{dz} \quad \frac{dE}{dt} \right] \qquad (1.17)$$

We will see later that this is consistent with momentum being seen as a spatial type of energy and thus no more than a different way of writing the standard definition.

11. *Electric Charge is Invariant under Change of Velocity:* Electric charge, say the charge of an electron, is invariant with velocity (the same at all velocities). If the electric charge of

electrons were different at different velocities, then electrons orbiting in different orbits would have different electric charges and the whole universe would fall to bits. This does not mean that the electric field of an electron is the same at all velocities. There are time dilation and length contraction effects that cause the electric field to change with velocity.

12. *The Unification of Electric Force with Magnetic Force:* Although electric charge is invariant under change of velocity, an electric field turns into an electric and magnetic field when it is moving. In special relativity, the electromagnetic field is expressed as a single 4-tensor. The electromagnetic 4-force is calculated by multiplying this 4-tensor by the 4-velocity and the (scalar) electric charge, q. When the spatial part of the 4-velocity vector is zero, we have pure electric field.

13. $E = mc^2$ *leads to QFT and we have to reject Quantum Mechanics:* One of the results of special relativity is that energy can be converted into mass and vice-versa. This means that particles like electrons can be created out of nothing more than energy or can be annihilated into energy. Thus, particles can be created and destroyed. The Schrödinger equation of quantum mechanics cannot handle this, and we have to reject quantum mechanics in favour of quantum field theory, QFT. It also means that an amount of energy has a mass equivalent to it.

14. *Mass Increase:* We tend to think of mass as inertial charge, but, unlike electric charge, mass increases with velocity. We have:

$$m' = m_0 \frac{1}{\sqrt{1 - \dfrac{v^2}{c^2}}} = m_0 \gamma \tag{1.18}$$

Where m_0 is the mass of an object in the "stationary world" as seen by the stationary observer and m' is the corresponding mass in the "moving world" also as seen by the stationary observer. It is not the rest-mass, m_0, that is increasing; it is the sum of the rest-mass and the mass equivalent to the kinetic energy that is increasing. We are used to the idea that a rapidly moving canon ball has more kinetic energy than a slowly moving canon ball of the same mass. Thus, given that energy is equivalent to mass, mass increase is not that mysterious. We emphasize: the rest mass, m_0, of a particle is the same at all velocities. Like the electric charge, the rest mass of a particle is invariant under change of velocity.

Aside: Mass can be converted into energy. Because of the energy needed to bind an atomic nucleus together (binding energy), the mass of an atomic nucleus is not equal to the sum of the masses of its parts. This does not happen with electric charge. The electric charge of an atomic nucleus is equal to the sum of the electric charges of its parts[6].

---

15   A. *Breakdown of Simultaneity:* Spatially separated events that are simultaneous for one observer are not simultaneous for another observer who is moving relative to the first observer. To put it another way, observers who are pointing in different directions in space-time do not agree on the simultaneity of spatially separated events.

B. *Events are Ordered:* Although two observers moving at different velocities will not agree that two spatially separated events are simultaneous, they will agree on the order of those events up to simultaneity – which was cause and which was effect. All observers at all velocities will agree that the first event happened first or was simultaneous with the second event (one velocity only) and that the second event happened second or was simultaneous with the first event (same one velocity). No observer will think that the second event happened first. This is because the rotations in space-time are restricted. (Roughly, and wrongly, but pedagogically picturesquely, to being only 90° out of the 360° we have in Euclidean space). This restriction is the finite limiting velocity (the speed of light). Without the limiting velocity, we would not have ordered events and cause and effect would be meaningless.

[6.] This is called the principle of superposition.

Thus, only observers moving in unison at one particular velocity will agree that two spatially separated events happened at the same time - simultaneously. If the two events are separated in time, then only observers moving in unison at zero velocity (stationary) will see two temporally separated events as happening at the same place. All other observers see two events that happen at different positions in space-time as being separated by some space and some time.

16. *The Unification of the Building Blocks of the Universe:* Within the Newtonian theory, space is a thing separate from time. In special relativity, these two separate things are combined into a single entity that we call space-time. The same is true of momentum and energy which are combined into momenergy; of force and power; of electricity and magnetism, which are combined into electromagnetism; and of current density and charge density. Instead of two conservation laws, conservation of momentum and conservation of energy, within special relativity, we have only the conservation of momenergy. There seems to be a problem here. Energy is measured in KgM^2S^{-2} whereas momentum is measured in $KgMS^{-1}$. The resolution of this seeming problem is that, within special relativity, length is taken to have the same mass-dimensions as time:

$$[L] = [T] \qquad (1.19)$$

This means that both energy and momentum are measured in kilograms – think $e = mc^2$. So, because space-time is one entity, space has the same mass-dimension as time; because space has the same mass-dimension as time, momentum is measured in the same units (mass-dimensions) as energy. It two things are measured in the same units (mass-dimensions), they must effectively be the same stuff. It is the unification of space with time that "causes" the unification of energy with momentum (and all the other unifications as well).

It is important to realize that a stationary observer viewing a physical system that is stationary with respect to her will always see the divided version of the unification. For the

stationary observer, there will always be both the conservation of energy and the conservation of momentum for
co-stationary systems.

17. *Special Relativity Approaches the Newtonian Theory at Low
Velocities:* Although Newtonian mechanics is not perfectly
accurate except when the velocity is zero, it is still very accurate, even at quite high velocities. NASA uses Newtonian
mechanics to calculate the trajectories of interplanetary
space missions because it is accurate enough at the velocities of these space missions and it is difficult to calculate with
general relativity. Newtonian mechanics is correct at zero
velocity, and so special relativity mechanics must, and does,
approach Newtonian mechanics, also known as the Newtonian limit, as the velocity approaches zero. This is expressed
by the mathematical fact that, when the velocity is zero,
gamma is unity:

$$\gamma_{v=0} = \frac{1}{\sqrt{1 - \frac{0^2}{c^2}}} = 1 \qquad (1.20)$$

EXERCISES

1. Draw a set of space-time axes such that the time axis is in the
horizontal direction and the space axis is in the vertical direction. Put two separate dots on the positive half of the time
axis. What is the spatial separation of the two dots? The dots
are separate in time but not in space. Super-impose another
set of space-time axes on to the drawing with the time axis at
$45°$ to the horizontal. What is the spatial separation of the two
dots in the new axes?

2. A photon of light is emitted from the big bang at the start of
the universe. 13.8 billion years later, an astronomy student
captures the photon in a telescope on Earth. Assuming that
the time dilation formula applies to photons of light:

a. How old is the universe as measured by the photon of light
(not the astronomy student)?

    b. How much time has the photon of light travelled through in its own view?

    c. How much space has the photon of light travelled through in its own view?

    d. What is the distance (interval) in space-time that the photon has travelled through in its own view?

    e. How much time has the photon travelled through in the astronomy student's view?

3. Does mass increase depend upon the direction of the velocity? Does time dilation depend upon the direction of the velocity? Does length contraction depend upon the direction of the velocity?

4. A distant galaxy of mass M_{Gal} is receding from the Earth at a velocity of $0.9c$. Assuming that the mass increase formula applies to distant galaxies, what is the mass of the distant galaxy as seen from Earth?

5. The mass-dimensions of energy are $[E] = \dfrac{ML^2}{T^2}$ (think $E = mc^2$). The mass-dimensions of force are $[F] = \dfrac{ML}{T^2}$ (think $F = ma$). What are the mass-dimensions of $\dfrac{dE}{dr}$?

6. A star with a single orbiting planet passes the Earth at $0.8c$. In the view of the star, its planet's orbit is perfectly circular. Use the length contraction formula to calculate the eccentricity of the planet's orbit as seen from Earth.

    Note: $e = \sqrt{1 - \dfrac{b^2}{a^2}}$.

7. There are three students traveling in the same direction. The biology student passes the physics student at $0.8c$. At the same time, the physics student passes the mathematics student at $0.9c$. At what velocity does the biology student pass the mathematics student?

8. a) What is the inner product of the two 4-vectors:
    $[4 \quad 1 \quad 1 \quad 2]$ & $[3 \quad 2 \quad 1 \quad 1]$?

b) What is the inner product of the two 4-vectors:
$[3 \quad 2 \quad 1 \quad 2] \, \& \, [3 \quad 2 \quad 2 \quad 2]$

9. A rocket moves perpendicularly away from the Earth at $0.5c$. As it does so, it accelerates at a rate which its own instruments declare to be $0.4c$ per second. What is the rocket's acceleration as seen by an Earth-bound observer?

10. A star moves away from Earth at $0.9c$. In its own reference frame, the star is burning 4×10^6 metric tons of mass per second (that is 4×10^6 of its own metric tons in one of its own seconds). What amount of mass is the star burning per second in the Earth's reference frame?

11. An art student traveling at $0.9c$ hits her thumb with a hammer. There is a stationary math student watching her. Given that force is not invariant with velocity, which student feels the most pain? How does time dilation affect this?

3

SPECIAL RELATIVITY IN PHYSICS

Special relativity is one theory of physics; general relativity is another completely different theory of physics. It is confusing that they both share the appellation "relativity". They were both developed by Albert Einstein. Special relativity deals with space in which there is no gravitational field. General relativity deals with a gravitational field. Special relativity is a universal theory in that there is no limit on the distance between observers – but see the cosmology chapter later. General relativity is a local theory in that it applies only over an infinitesimally small area of the universe. In the original German, Einstein initially called special relativity "Invariant theory", which would have avoided the confusion.

We inhabit only one universe. Clearly, this only one universe ought to be explained by only one theory. Rather embarrassingly, modern physics has two theories. Those two theories are quantum field theory (QFT) and general relativity. Between them, these two theories explain the four forces of nature. QFT is the theory of particle physics that covers the strong, weak, and electromagnetic forces. General relativity is a theory of the gravitational force. QFT contains within it special relativity. Special relativity has been completely unified with particle physics into QFT. There is no unification of general relativity and QFT. The two theories of QFT and general relativity are philosophically and mathematically distinct. It is remarkable that both these two very different theories are both extremely accurate descriptions of reality; their correctness has been verified by many precise experiments. In spite of the fundamental difference of QFT and general relativity, much theoretical work has been, and is being, undertaken to try and unify these two theories together into one grand unified theory that describes the one real universe. String theory's *raison d'etre* is this quest for the unification of QFT and general relativity.

We seek to build the universe in such a way that the physical laws of that universe are invariant under change of velocity. We call this Lorentz invariant physics. To do this, we need the Lorentz transformation; this is a convoluted way of saying we need the space-time rotation matrix. We have that and will reveal it shortly. To build the universe, we also need 4-tensors, which we will meet later. There are three types of 4-tensor that we need; these are scalars (rank zero 4-tensors), 4-vectors (rank one 4-tensors), and rank two 4-tensors. These mathematical objects are enough to build special relativity.

3.1 THE HISTORY OF SPECIAL RELATIVITY

In 1861–62, James Clerk Maxwell (1831–1879) unified electricity and magnetism into electromagnetism. One of the consequences of this unification is the wave equations of the electric and magnetic fields:

$$\nabla^2 B = \varepsilon_0 \mu_0 \frac{\partial^2 B}{\partial t^2} \quad : \quad \nabla^2 E = \varepsilon_0 \mu_0 \frac{\partial^2 E}{\partial t^2} \tag{3.1}$$

From these, it is deduced that the velocity at which these electromagnetic waves propagate is given by:

$$c = \frac{1}{\sqrt{\varepsilon_0 \mu_0}} \sim 300,000 \text{ km/sec} \tag{3.2}$$

This equation tells us that electromagnetic waves propagate at the speed of light, but it does not tell us to what that speed is relative. It left the Victorian physicists with questions:

1. Does light travel at 300,000 km/sec relative to the emitting source or relative to the receiving observer or relative to the center of the universe or relative to what?

2. Further, if light is a wave, it must surely need an undulating medium to carry it. What is this medium? Surely, empty space is no such medium.

Experiments had convinced physicists that the speed of light was a physical constant. However, the idea that all observers, including ones that were in relative motion to each other, should each measure the speed of the same beam of light and each get the same answer seemed ridiculous because it meant losing the relative velocity of the observers. Literally, the sums did not add properly!

Thus, without being able to find a shred of evidence to substantiate the idea, physicists felt themselves forced to invent a luminiferous aether that filled empty space. The aether was postulated to be the undulatory medium that carried light waves. Think about it; everyone knows that waves are an undulation in a medium; to propagate, electromagnetic waves must have a medium. Is there anyone out there who can explain to your author how waves propagate without a medium in which to undulate? Given Maxwell's calculation of the velocity of light, the universe needed a reference frame against which to measure this velocity. If there was an undulatory medium for light waves, then this medium would provide a fixed reference frame against which the speed of light could be measured and against which it would be constant. Even if waves could propagate without a medium, surely this reference frame was needed. Indeed, one can interpret Maxwell's calculation as being the prediction of the existence of such a reference frame. The aether was postulated to be the reference frame relative to which the velocity of light was the constant 300,000 kilometers per second.

This aether was strange stuff. It provided a medium for the propagation of light waves, and empty space was full of it; yet empty space was still empty. The aether weighed nothing. It did not gravitate. It did not shine. In fact, it was completely undetectable. Even so, it had to exist and, seemingly, had been predicted to exist.

It is easy for we with hindsight to decry the idea of the aether, but put yourself into the shoes of the physicists of the time. Would water waves exist if there was no water. To repeat, a wave, after all, is an undulation of a medium; it is surely nonsense to think a wave could exist without a medium in which to undulate, and in 1887

Heinrich Hertz (1857–1894) detected the electromagnetic waves traveling through space. Of course, today, we know that the vacuum is a seething mass of virtual particles. Is this not just another version of the aether?

Aside: The orbital speed of the Earth around the sun is circa 30 kilometers per second.

The orbital speed of the sun around the center of the Milky Way is circa 300 kilometers per second. Thus, if the aether existed, the Earth would plough through it at a goodly rate.

In 1887, Albert Michelson (1852–1931) and Edward Morley (1838–1923) attempted to measure the Earth's motion through the aether. Unless the aether was fixed to the Earth (an unlikely and ugly scenario), the Earth would move through the aether. Since the velocity of light was presumed to be measured against the aether, by measuring the velocity of light in different directions, one could detect the Earth's motion through the aether as being in the direction in which the velocity of light was the least. In practice, it was possible to measure only a difference in the velocity of light in different directions using interference, but this would be sufficient to detect the aether. Not only did Michelson and Morely compare this the velocity of light relative to the Earth in different directions, but they repeated the measurements several times throughout the year as the Earth orbited around the sun thereby changing its direction through the aether. This was known as the Michelson-Morley experiment.

Michelson & Morley did not detect any difference in the velocity of light in different directions or at different times of the year, and so they did not detect the aether, and today we know that it does not exist, or, at least, it cannot be detected. It certainly is not a fixed reference frame against which we can measure the speed of light. Modern versions of Michelson and Morley's experiment[1] to detect the aether have confirmed its non-existence to an accuracy of one part in 10^{15}. The modern GPS navigation system itself proves to great accuracy that the aether does not exist because the

---

[1] A. Brillet & J. L. Hall Phys. Rev. Lett. 42 549 (1979).

GPS system relies on accurately timed electromagnetic communication whose timing must be the same at any time of the year. Similarly, the "Temps Atomique International" international atomic time relies on accurately timed electromagnetic communication whose timing must be the same at all times. Thus, not only do light waves seemingly not require an undulatory medium through which to propagate, but the speed of light is 300,000 km/sec for all observers.

In 1905, following the enunciation of the relativity principle by Poincare (1854–1912) in 1904, Albert Einstein, accepting the findings of Michelson and Morley at face value, started from the simple statement that the speed of light is the same for all observers, including observers in relative motion to each other. From this, he deduced the special theory of relativity. He published his work in the paper "Zur electrodynamik bewegter korper" in the annallen der physic 17 891. In English, the paper is called "On the electro-dynamics of moving bodies". The mathematics of the special theory of relativity includes the expression:

$$\gamma = \frac{1}{\sqrt{1 - \dfrac{v^2}{c^2}}} \tag{3.3}$$

This had previously been written by both George Fitzgerald (1851–1901) and Hendrik Lorentz[2] (1853–1928) as a means of explaining the failure of Michelson and Morley to detect the aether. It is known as the Lorentz-Fitzgerald contraction, or, more often, as just the Lorentz contraction. Mathematically, Fitzgerald and Lorentz were very close to special relativity; conceptually, they were a universe away from it.

Note that:

$$\gamma \geq 1 \tag{3.4}$$

at all velocities greater than zero and less than c.

In 1907, Hermann Minkowski (1863–1909), a former tutor of Albert Einstein, re-wrote Einstein's special theory of relativity as

---

[2.] Einstein said of Lorentz, "I admire this man as no other".

4-dimensional space-time in which change of velocity is just rotation in space and in which invariance of the laws of physics under change of velocity was no more than the isotropy of space-time. It was Minkowski that unified space and time into a single space-time entity, and it is Minkowski's work that is the basis of this book, albeit that we also use matrices where Minkowski used only 4-vectors. It was in his address to the 80th Assembly of German Natural Scientists and Physicians in 1908 that Minkowski declared, the now famous, words, "… Henceforth, space by itself, and time by itself, are doomed to fade away into mere shadows, and only a kind of union of the two…". It is often the case that a physical theory is first roughly hacked from experimental results by a genius who sees the theory within the results as a sculptor sees the statue within the boulder. Most often, the genius leaves the theory rough and blemished, and it is for others to come along and do the laborious polishing. So it was that Einstein's rough hacking was turned into a beautiful theory by Minkowski. 4-dimensional space-time is now often referred to as Minkowski space-time in honor of him.

In 1912, Ludwik Silberstein (1872–1948) published work in which he used complexified quaternions (bi-quaternions) to rewrite the theory of special relativity[3]. Silberstein did much to promote the teaching of special relativity in universities.

In 1925, Paul Dirac (1902–1984) combined special relativity and quantum theory in his famous Dirac equation and founded Quantum Field theory. Dirac's deduction of the existence of anti-matter derives from special relativity, as does Sommerfield's fine structure of atomic spectra, and Pauli's explanation of the connection between spin and statistics.

Today, following Minkowski, we see special relativity as no more than the isotropy of space-time. The invariance of the speed of light is a useful pedagogic tool to us, but it is not a central part of special relativity; it is merely a consequence. None-the-less the invariance of the speed of light was a *force majeure* that drove physics to the theory of special relativity.

[3.] Phil. Mag. S. 6, Vol. 23, No. 137 (May 1912), 790–809.

Prior to the twentieth century, there were two failings of Newtonian mechanics. The first was a 43 seconds of arc error in the predicted orbit of the planet Mercury[4] which special relativity did not correct. (It was general relativity which, in 1915, solved this problem.) The second failing of Newtonian physics was that Maxwell's theory of electromagnetism was not invariant under Newtonian transformations. The electromagnetic force, given by $\vec{F} = \vec{E} + \vec{v} \times \vec{B}$, increases with velocity, and Newtonian forces do not change with velocity. It was this that drove the need for a mechanics to replace Newtonian mechanics. The advent of special relativity was the advent of that new type of mechanics, and it solved the problem with Maxwell's theory of electromagnetism. By 1925, the new mechanics had replaced Newtonian mechanics in theoretical physics even though, except for very high energy particle physics, Newtonian mechanics is still used today for all practical physics and engineering.

Relativity, both the special theory and the general theory, were not taught as part of an undergraduate degree until the 1950s, and not widely so then. It is interesting to speculate why. Was it because the established individuals in the universities did not believe in relativity? There is some, but not much, evidence of this. Was it because the established individuals in the universities could not understand relativity? There is some, but not much, evidence of this. Was it because the established individuals in the universities were too lazy to bother learning relativity – you can't teach old dogs new tricks? There is some, but not much, evidence of this. Was it because physics was primarily studied for practical purposes in most universities and relativity did not seem to have any practical consequences? It is your author's opinion that this was the main reason, and this is a good reason. Practical things feed people! None-the-less, it seems always to take a generation for science to take a major step forward, and it does appear as if the old wood needs to be pruned out to let the young wood bear fruit.

[4.] The 43 seconds or arc error was first detected by Leverrier in 1859. It was postulated that this error was caused by the presence of a yet undiscovered planet nearer to the sun than Mercury. The postulated planet was named Vulcan, which is where Mr. Spock comes from.

Could special relativity have been discovered earlier than 1905? It was certainly an opinion of Lord Cherwell (1886–1957) that, "… if scientists had had their wits about them, they ought to have been able to reach relativity theory by pure logic soon after Isaac Newton, and not have to wait for the stimulus given to them by certain empirical observations that were inconsistent with the classical theory"[5]. Historians of science will give a hundred similar examples of the unnecessary slowness of progress. It was the opinion of the mathematician Carl Friedrich Gauss (1777–1855) that scientific theories are like flowers that do not bloom until their springtime has come and then all related theories bloom together. In 1905, both special relativity and quantum mechanics sprouted, and both were gardened by Einstein. Perhaps 1905 was the Gaussian springtime for which special relativity waited.

The invention of the aether is an example to us of how easily we can mislead ourselves, but it made perfect sense at the time. People felt that there could be no other explanation. The physics text-books of late Victorian times are full of calculations of vortices and swirls within the aether that explain the result of the Michelson Morley experiment. In 1927, there were still conferences discussing the aether, and, in 1928, Lorentz himself died still believing in the aether. Physicists will not let that kind of thing happen to them again. They wouldn't believe in virtual particles if such things did not really exist, would they? They wouldn't believe in higher dimensional spaces curled up into extremely small tubes if such things did not really exist, would they? Can higher dimensions be curled up as string theorists tell us or is this an even bigger whopper than the aether?

What tangled webs we achieve
In the theories we conceive

---

[5] Lucas & Hodgson. Space-time and electromagnetism. pg. 293.

CHAPTER 4

THE NATURE OF SPACE

Humankind does not properly understand the nature of empty space. Humankind has observed only one type of space[1]. We seem to live in, and thus have experience of, what seems to be only one type of space. This is the 4-dimensional space-time in which your author is currently sitting. Humankind has never observed a 5-dimensional space or a 3-dimensional space. Humankind has never observed a spatial (or temporal) dimension attached to our 4-dimensional space-time to form 5-dimensional space, and humankind has never observed a dimension being ripped from this 4-dimensional space-time to form 3-dimensional space. Indeed, based on observations, it seems that dimensions cannot be ripped from, or added to, space-time. Many books tell us that a flat plane of zero thickness is 2-dimensional space, but no one has ever actually seen a flat plane of zero thickness. It is a product of the mathematician's imagination. The same books will describe a spatial volume as being 3-dimensional, but no one has ever stopped time to see if this is correct, and it seems that it is not possible to stop time (that is to set time to zero). Nor do we know from observation whether our 4-dimensional space-time is four identical 1-dimensional spaces fixed together, two

---

[1] It is possible, and your author thinks sensible, to interpret quantum field theory as being based on symmetries in different types of "internal" spaces. Thus, it might be inferred that humankind has observed these different types of space, based on unitary Lie groups, in particle physics.

2-dimensional spaces fixed together, one 3-dimensional space and one 1-dimensional space fixed together (as opposed to Newton's view that space and time were not fixed together), or one 4-dimensional space that stands alone. Indeed, might it be several types of 4-dimensional space laying over each other and sharing the same axes? We actually know very little about space.[2] Most of what we think we know is no more than the invention of mathematicians.

If 4-dimensional empty space is just four 1-dimensional spaces fitted together, from where do angles come and why is time different from space? Why do we observe two types of trigonometric functions (the Euclidean, {cos( ), sin( )}, and the hyperbolic, {cosh( ), sinh( )}) in our 4-dimensional space-time.

Aside: Lie group theory is about rotations in different types of space. The types of space it considers are formed from n copies of either $\mathbb{R}$ (called the orthogonal groups) or n copies of $\mathbb{C}$ (called the unitary groups) or n copies of the quaternions, $\mathbb{H}$, (called the symplectic groups) or five oddball spaces formed from copies of the octonians.

4.1 POSSIBLE TYPES OF DISTANCE FUNCTIONS

A normed algebra is a type of numbers (like the real numbers, $\mathbb{R}$, or the complex numbers, ($\mathbb{C}$) together with the concept of the "length" of a number. The "length" of the number is called its norm. In the real numbers, this concept of length (norm) is just the value of the real number – its distance from the origin; in the complex numbers, $\mathbb{C}$, this concept of length is the modulus of the complex number – its distance from the origin. A normed algebra is thought of as being a type of empty space. The concept of length in a normed algebra brings with it the requirement that the norm of a product of two numbers be of the same form as the norm of each of the two numbers – we calculate the "length" of each number with the same form of mathematical expression. In the complex numbers, $a + ib$, this expression is $Norm = a^2 + b^2$.

[2.] See: Dennis Morris : *The Physics of Empty Space* – ISBN: 978-1507707005

With a 3-dimensional quadratic distance function (norm), it is not possible to form a normed algebra in which the form of the norm of a product of two numbers is the same as the form of the norms of the two numbers. It is possible for such algebras to exist in two dimensions and in four dimensions. This is because:

$$(a^2 + b^2)(c^2 + d^2) = (ac - bd)^2 + (ad + bc)^2 \equiv X^2 + Y^2 \quad (4.1)$$

And

$$(t^2 + x^2 + y^2 + z^2)(a^2 + b^2 + c^2 + d^2)$$
$$= (at - bx - cy - dz)^2 + (ax + bt + cz - dy)^2 \quad (4.2)$$
$$+ (ay - bz + ct + dx)^2 + (az + by - cx + dt)^2$$
$$\equiv W^2 + X^2 + Y^2 + Z^2$$

But

$$(a^2 + b^2 + c^2)(e^2 + f^2 + g^2) \neq X^2 + Y^2 + Z^2 \quad (4.3)$$

Aside: The reader might want to try setting one of the variables to zero in the 4-dimensional expression above. The reader will then see that, in 3-dimensions, this does not give a normed algebra. These algebraic facts are based on the determinants of multiplicatively closed sets of matrices derived from finite groups – see later. Ultimately, 3-dimensional algebras with quadratic norms are impossible because order four groups do not have order three sub-groups. In 3-dimensions, we have (based on the group C_3):

$$(a^3 + b^3 + c^3 - 3abc)(d^3 + e^3 + f^3 - 3def) \equiv X^3 + Y^3 + Z^3 - 3XYZ \quad (4.4)$$

Of particular interest to us is:

$$(a^2 - b^2 - c^2 - d^2)(e^2 - f^2 - g^2 - h^2) \neq W^2 - X^2 - Y^2 - Z^2 \quad (4.5)$$

Notice the not-equal-to sign. Two rotation matrices multiplied together make a rotation matrix. A rotation matrix is a rotation because it holds invariant the distance function (from the origin). The only multiplicative invariant (a thing that is preserved by multiplication) of a matrix is the determinant of that matrix. Thus, the form of the distance function of a space is the form of the determinant of the rotation matrix. For any two matrices, we have $\det(A)\det(B) = \det(AB)$. Because of (4.5), there cannot be a proper 4-dimensional rotation matrix in a space-time with this distance function. Which means that the space-time in which we seem to sit is not really a space - Hm!

We can do 2-dimensional rotations in the space in which we sit, but we cannot do a 4-dimensional (or a 3-dimensional) rotation in this space[3]. It makes you think that we are sitting in a mixture of 2-dimensional spaces rather than in a single 4-dimensional space, but why do these spaces not collapse on to each other? What holds them in the 4-dimensional form we observe? Later in this book, we will see that modern physics takes the view that 4-dimensional space-time is represented by the Lorentz group which is a seen as a set of (commutation relations between) 2-dimensional rotations. In spite of the Lorentz group being based on a set of 2-dimensional rotations, it contains 4-dimensional rotations in which the commutator of two 2-dimensional space-time rotations (also called boosts) is not another space-time rotation but is a purely spatial rotation. This is usually phrased as "the commutator of two boosts is a rotation'.

4.2   WHAT IS EMPTY SPACE?

Empty space is amazing stuff. It is widely believed to be nothingness. We take empty space to be empty; we take it to be nothing. Empty space has nothing in it and it is nothing itself. Yet, still, amazingly, this nothingness has properties.

Magnitude: There is more empty space between the planet Earth and the planet Pluto than there is between the planet Earth and the planet Mars. How can there be more of nothing? Presumably, there are just more virtual particles between us and Pluto than there are between us and Mars.

Homogeneity: Empty space is the same everywhere. The properties of this nothingness are the same everywhere. Of course, the concept of "everywhere" is a concept associated with space.

Isotropy: Empty space is the same in all directions.

[3.] Actually, as well as 2-dimensional rotations, there are 4-dimensional rotations in our space, but the complications are such that we have not space for them here. We deal with them much later in this book when we look at the A_3 algebras and the Lorentz group.

Dimension: It seems that the empty space of our universe is 4-dimensional (3 spatial dimensions and one time dimension). If empty space is nothing, how can it have dimension? We will later meet spaces of dimension other than four.

The central understanding of physics is that physical laws are invariant under certain changes such as rotation in space-time. Ought not one of these changes be the dimensionality of the space? It seems not. If a central force is proportional to r^{-n}, then circular orbits (of planets say) are stable[4] only if $n < 3$. There are symmetries in Maxwell's equations and in Dirac's equations of the electron that are not present in space-times of dimension other than four[5].

Electrical permittivity and magnetic permeability: Empty space allows electromagnetic fields to penetrate it with a particular strength to distance ratio. It is the value of this ratio that determines the particular value of the speed of light. So, why that particular value? Why not half of that value, or why not a random value? We will see later that light travels at exactly the speed that is necessary to keep events in order in the universe and therefore keep cause and effect. So, it is the values of electrical permittivity and magnetic permeability that "cause" cause and effect to exist – that's weird!

Different types: Empty space comes in different types – this alone is utterly amazing. Special relativity is about space-time. Distance in 2-dimensional space-time is calculated as $d = \sqrt{t^2 - z^2}$, while distance in 2-dimensional Euclidean space is calculated as $d = \sqrt{x^2 + y^2}$. We wonder why one of the dimensions in space-time is a time dimension and the other one is a space dimension. Time flows inexorably forward, but it is easy to be stationary in space. We observe that $x^2 + y^2 = y^2 + x^2$ but that $t^2 - z^2 \neq z^2 - t^2$; perhaps this is why the two dimensions of space-time are different from each other. More likely, the difference between the cosh( ) function and the sinh( ) function is the root of the difference between space and time. Later in this book, we will introduce the reader to other different kinds of empty space. Empty space comes in different types - there are different types of nothingness!

---

[4] Dynamics by H. Lambe Cambridge University Press (1914) pp. 256–258.
[5] C.Lanczos "The splitting of the Riemann tensor" Reviews of Modern Physics, 34, 379–389 (1962).

Extent: Empty space seems to be infinite. If it was finite, would there be something outside of it?

Fields of empty space hold momentum and energy: Picture a magnet that acts at a distance through empty space to attract a ferrous object. It does not extend a physical hand to grab the object or touch it in anyway, and yet it still attracts it. We invent the concept of magnetic field[6] (and electric field and gravitational field) to "explain" the magnetic attraction, but what is a field? A field is just as mysterious a concept as is the "action at a distance" that it was invented to explain. The concept of field explains nothing, but we find that fields can hold momentum and energy. If we release two electrically charged objects into space, they accelerate towards (or away from) each other; from where comes the kinetic energy? Fields are no more than empty space – nothingness – yet they hold energy.

Knowledge of rotation: The Earth spins, and, as a result of that spinning, it bulges at the equator. How does the Earth know that it is spinning, and, hence, that it should bulge? Why does the Earth not think that the universe about it is spinning while the Earth itself does not spin? Imagine a universe containing nothing more than two buckets of water. One of the buckets is spinning clockwise; or is it the other bucket that is spinning counterclockwise? The surface of the water of the spinning bucket will be curved by the centrifugal force associated with the spinning. How does the water in the spinning bucket know that it is in the spinning bucket rather than in the stationary bucket and thus curve its surface? Empty space seems to know that the Earth is spinning.

Motion: Empty space allows objects to move relative to each other at a particular velocity – that is the ratio of a particular amount of space to a particular amount of time. This is just a particular direction in space-time, but why does space have direction? Why doesn't the universe just stay motionless?

Knowledge of inertia: Empty space "feels" acceleration just like it feels rotation. Empty space does not "feel" velocity; why should it "feel" acceleration?

Orientabilty: Within empty space, we can tell the difference between a left hand and a right hand. We cannot manoeuvre one

---

[6.] Actually, it was Michael Faraday (1791–1867) who invented the concept of a field.

into the shape of the other. Nothingness, that is empty space, has this permanent property that prevents us from turning a right-hand into a left hand.

Locality: If it were possible to move at infinite speed, the difference in time between being in one place and being in lots of other places would be zero. An object would be in all these places at the same time, or, in two places at once, if you prefer. Because our universe has a limiting finite velocity (the speed of light), an object cannot be in two places at the same time and we have the phenomenon of locality. So, empty space has a structure. This structure becomes more apparent in the particular nature of the trigonometric functions of a particular space.

Linearity: All the types of space we meet in this book, and all the basic types of space that are within mathematics have the property of linearity. Linearity means that when we rotate a straight line, it remains straight rather than becoming curved and its length remains unchanged. This means that, if one observer sees an object to be moving without acceleration, another observer moving without acceleration relative to the first observer will also see that object to be moving without acceleration. This is important if we are going to insist on the invariance of a physical system under rotation in spacetime, or any other type of space. Matrices are often referred to as linear algebra because they are the algebraic expression of movement in linear space. Normal differentiation is also a linear operation.

> **NOTE**
>
> *A linear space can be thought of as a "flat" space rather than a "curved" space. General relativity is nonlinear. We often think of the space of general relativity as being "curved" space, and we need to use a special type of differentiation (covariant differentiation) to work with general relativity.*

4.3 VIEWS OF THE NATURE OF SPACE

Poincare's view: The physicist is free to ascribe to physical space any one of a number of mathematically possible geometric structures provided she makes suitable adjustments to the laws of mechanics and optics and to the rules for measuring length[7]. Of the possible geometries, Poincare thought that we should choose a Euclidean geometry because it is the simplest. Good idea! but then Einstein produced general relativity, which uses non-Euclidean geometry and which is simpler than the Euclidean geometry. Poincare's view is that there are two ways to construct physics; one way is to specify the rules for measuring length and angle (the nature of space) and then adjust the physical laws to fit with these rules; the other way is to specify the physical laws and then adjust the nature of space to fit with these laws. General relativity adjusts the nature of space to fit the physical law of gravity. Newton adjusted the law of gravity to fit his view of space. To Newton, space and time are uniform and so a body in a gravitational field changes its velocity - it accelerates. To Einstein, a body in a gravitational field moves at a uniform velocity (a free-fall geodesic) but space and time are not uniform – space is "squashed up" and this appears as acceleration of the body moving through it.

Poincare's view asks whether it is the task of physics to single out the one type of geometric space that puts physical laws in their simplest form or is it the task of physics to single out the physical laws that put space into its simplest form? What if space is a phenomenon of electromagnetism and not a thing separate from the physical laws of the universe?

Mach's view: Would empty space exist if there were no objects in it to mark its extent? It was the opinion of Ernst Mach (1838–1916) that it was the nature of empty space that empty space would not exist if objects did not exist to mark its extent. He opined that empty space was just a set of relations (distances apart) between objects. He opined that rotation and inertia were measured against the average motion and average inertia of objects in the universe. If there were no objects in the universe, there would be nothing against which to specify zero acceleration and zero rotation and thus no such thing as rotation or a straight line – straight lines are the paths followed by inertial bodies. With no straight lines, there is no space. The idea was that an object somehow "knows" about every other object in the

---

[7.] The Philosophy of Space & Time – Hans Reichenbach.

universe and measures its inertia and rotation against all those other objects. Mach pointed out that if I stand up and spin around, two things happen; the stars above me rotate relative to me and my arms lift due to centrifugal force; the two things must be connected, but how? In general, Mach took the view that space is not a thing and that empty space does not exist in its own right.

If we are to reject Mach's view of space, how do we explain the seemingly absolute nature of acceleration and rotation? It is often said within texts on special relativity that Einstein did away with Newton's absolute space. Einstein did away with absolute velocity; he did not do away with absolute acceleration and absolute rotation. Unless we accept a Machian type of view, we still have to accept absolute space, all-be-it without absolute velocity.

Mach's view of space was and still is taken very seriously by theoretical physicists. On one hand, Ozvath and Schucking have shown that general relativity fails to exclude solutions that contradict Mach's view[8]. On the other hand, Dicke[9] and Brans[10] have shown that by treating gravitation as a scalar-tensor field (general relativity treats gravity as a tensor field), Mach's view is fully incorporated into the theory of gravity. Further, in 1970, Solomon Schwebel formulated a relational theory of mechanics and conservation laws without any reference to absolute space[11]. Yet further, it has been discovered[12] that there are also other problems with non-Machian views.

Kant's view: Immanuel Kant (1724–1804) wrote "Critique of Pure Reason', and his fame is as a philosopher[13]. None-the-less, he was the first to suggest that the Milky Way (*via lactea*) galaxy was rotating[14], and he had views upon the nature of empty space. He was

---

[8.] I. Ozvath & E. Schucking *"Finite rotating universes"* Nature 193 1168–1169 (1962).

[9.] R.H.Dicke *"Mach's principle and equivalence"* in Evidence for gravitational theories – Academic press, New York, 1962.

[10.] C.Brans & R.H.Dicke *"Mach's principle and a relativistic theory of gravitation'*, Physical review 124, 925–935 (1961).

[11.] S.L.Schwebel "Mach's principle and Newtonian mechanics" *International Journal of theoretical physics*, 3, 145–152 (1970).

[12.] J.Stachel "Einstein's search for general covariance" – Pub. in Einstein and the history of general relativity, Birkhauser, Boston. pp. 62–100 (1989).

[13.] He also wrote the much less famous "Critique of Practical Reason'.

[14.] Theory of the Heavens, Immanuel Kant (1755).

of the view that, although empty space exists *"a priori"*, what we observe is the product of the way our perceptions and our minds are "wired up" and that, if our minds (brains) were "wired up" in a different way, or if we perceived the universe through different means, then we would perceive a different type of empty space. Today's philosophers are more likely to take the view that our mind is manifestation of our brain and that our brain exists in empty space. None-the-less, Kant unsettles our feelings of certainty about space and time. We observe that we live in a universe that has three spatial dimensions and one time dimension, but can we trust our observations. Is it possible that our minds are deceiving us and that space is something other than as it appears?

Aside: Animal psychologists tell us that turkeys have no sense of time. Apparently, turkeys live their entire lives in the present. Seemingly, a sense of time is associated with the mammalian part of the brain. This must be why turkeys do not look forward to Christmas.

We simply assume that what we see is empty space because it seems that way. Perhaps what we observe is several spaces all sharing the same axes, and they add together to form what we see – see later.[15]

The conventional mathematician's view: Within mathematics, the usual view of empty space is that it is n copies of the real numbers, $\mathbb{R}$, fixed together to form a n-dimensional space. On to this construction, the mathematician installs a distance function (often of a quadratic form) and a concept of angle (often with trigonometric functions copied from the complex plane, $\mathbb{C}$). We will not adopt this view in this book.

Aside: Metric spaces are viewed as being geometric spaces with a distance function. The mathematician may choose any function to be the distance function provided that it satisfies three requirements:

  i.   The distance from point A to itself is zero
  ii. The distance from A to B is the same as the distance from B to A

[15.] See: Dennis Morris : The Physics of Empty Space – ISBN: 978-1507707005

    iii.   The distance directly from A to B is less than or equal to the distance from A to B via C – the triangle inequality.

Space-time is not a metric space[16] and we have no interest in them.

Within a metric space, all of the nature of that space is "inside" the distance function because that is the only thing from which the space is constructed. Of course, this does not include the concept of angle. It is possible to construct a space from nothing more than 2-dimensional rotations, and we will see this when we look at the Lorentz group towards the end of this book. However, it is not obvious from where the distance function of such a space derives.

Aside: Riemannian spaces are metric spaces with a distance function that is a sum (plus or minus) of squared variables (think Pythagoras). These spaces are such that, by setting a variable to zero, we can "rip away" one of the dimensions or by adding another variable we can "stick on" another dimension. General relativity and string theory are both formulated in Riemannian type spaces[17].

The view taken by this book: In this book, the theory of special relativity will be deduced from the finite group C_2, and the book will go on to deduce electromagnetism and something similar to the Lorentz group and the space in which we sit from the finite group $C_2 \times C_2$. This book will not use $\mathbb{R}^n$ as the concept of space. This book disagrees with Mach, and we take it that space is a thing that exists in its own right. This book assumes that space is like numbers. Empty space truly is amazing stuff, whether it does or does not really exist, but it is very similar to numbers, which also do or do not really exist[18]. A mathematics student cannot drop a number upon her toe, no matter how big the number (or how big her toe). She cannot eat it nor do anything materially useful with it. However, numbers have magnitude. They are also homogeneous, and they have dimension. The complex numbers are 2-dimensional numbers, whereas the

---

[16.] Technically, space-time is not a metric space, but it is often spoken of as being a metric space because it is associated with a metric tensor.

[17.] Technically, space-time is not a Riemannian space because it has minus signs in the distance function.

[18.] Perhaps we need to expand our concept of existence.

real numbers are 1-dimensional numbers, and the quaternions are 4-dimensional numbers. Within these dimensions, numbers are isotropic. Numbers come in different types ($\mathbb{R}$, $\mathbb{C}$, $\mathbb{S}$, $\mathbb{H}$,...), as we will see in due course, and they are infinite in extent. The quaternions and anti-quaternions are oriented differently. We can even lay one type of number over another type and produce what appears to be a field by comparing the different norms (lengths) of those different types of numbers.

Numbers have many, and, in your author's opinion, all, of the same properties as empty space. We see this in that mathematicians will say that 1-dimensional space is isomorphic to the real numbers, $\mathbb{R}$, and that 2-dimensional Euclidean space is isomorphic to the complex numbers, $\mathbb{C}$. We will soon meet the hyperbolic complex numbers, $\mathbb{S}$. These are isomorphic to 2-dimensional spacetime. Mathematicians insist that types of empty space and types of numbers are not the same things even though they do have all the same properties. However, this book holds the opinion that types of numbers and types of empty space are the same thing, and that this means that we can understand empty space if we understand numbers. This book's view is that the different types of empty space correspond to different types of numbers, and that these different types of numbers "grow" out of the different finite groups (which we will introduce shortly). In the conventional view of space, each axis is a real axis, $\mathbb{R}^1$. In this book's view, there is one real axis in a space and the other axes are imaginary axes. The complex plane, $\mathbb{C}$, is an example of this kind of space. The quaternion space, which has one real axis and three imaginary axes, is another example.

It is remarkable that the theory of special relativity, and Maxwell's electromagnetism, and, subject to interpretation, the Lorentz group which represents the space we observe around us, can be derived from nothing more than the existence of the finite groups $\{C_1, C_2, C_2 \times C_2\}$ and the real numbers. C_1 is the foundation of the real numbers, which is the only 1-dimensional space[19]. C_2 is the foundation of the hyperbolic complex numbers, $\mathbb{S}$, which is space-time, and C_2 is also the foundation of the complex numbers, $\mathbb{C}$, which is 2-dimensional Euclidean space. C_2 is obviously the foundation of $C_2 \times C_2$.

[19.] C_1 is a sub-group of C_2, of course.

This book will base special relativity in the kind of space that derives from the finite group C_2. We will make no prior assumptions that space is formed from n copies of $\mathbb{R}$ fixed together, and we do not need, nor wish to use, this construction of space. The theory of special relativity is an understanding of the nature of empty space. In this book's view, it is also an understanding of the nature of the type of numbers that derive from the finite group C_2.

So, to sum-up the important paragraphs: We will deduce special relativity from the finite group C_2. We will not use $\mathbb{R}^n$ as our concept of space.

EXERCISES

1. Why is space-time, with distance function $d = \sqrt{t^2 - z^2}$ not a metric space?

2. The Lorentz group is effectively a set of six 2-dimensional rotation matrices. Three of these are spatial rotations and three are space-time rotations. Is this a valid way to describe the 4-dimensional space in which we seem to live? Do we need a distance function to fully specify a type of space?

3. Would it be more sensible to specify a space by a rotation matrix rather than a distance function?

CHAPTER 5

PHYSICAL CONSTANTS

As we study the physical nature of the universe, we discover that there are particular numbers that enter our equations. For example, the equation $e = mc^2$ includes the number $c = 299, 792, 458$, which is the speed of light. Other examples are the gravitational constant in $F = -\dfrac{GMm}{r^2}$, or the charge of the electron, or the mass of the electron.

Aside: In 1967, the General Conference of Weights and Measures (CGPM-1967) defined the second as "...the duration of 9,192,631,770 periods of the radiation corresponding to the transition between the two hyperfine levels of the ground state of the caesium-133 atom." In 1983, the CGPM-1983 defined the meter as the distance travelled by light in a vacuum in a time interval of $\dfrac{1}{299,792,458}$ of a second. Thus, the speed of light, by definition, is exactly 299,792,458 meters per second. Thus the speed of light is set by definition, and the length of the meter is measured against the speed of light. This is the opposite to the meter being set by definition and the speed of light being measured against the length of the meter.

In particle physics, we have a theory called the standard model that describes the physics of atomic particles. The standard model is the central understanding that we have of physics at very small distances (inside of protons). The standard model requires nineteen numbers (parameters) to be fed into it before it works. These are numbers like the charge of the electron – physical constants.

This book is about special relativity. It is central to special relativity that physical constants, as with all laws of physics, are the same regardless of the direction in which the measuring apparatus points in space-time – that is regardless of the velocity of the measuring apparatus. The invariance of the physical constant that is the speed of light is most often associated with special relativity, but the invariance of all such physical constants is included in special relativity.

Why these physical constants have the values that they do have is one of the central mysteries of modern physics. True, the actual values change with the size of the units (meters, seconds, coulomb…) that we use, and we could set the speed of light equal to 1 by changing the length of the meter or the duration of the second, but this does not explain why there are such numbers or why these numbers have the values relative to each other that they do. Why should the electron be the mass that it is? Why should gravity be the strength it is and not a million times stronger? Where do these physical constants come from?

Aside: Theoretical physicists often set Planck's constant and the speed of light equal to unity, $\hbar = c = 1$. The mass-dimensions of these constants are:

$$[c] = \frac{L}{T} \qquad\qquad [\hbar] = M\frac{L^2}{T} \qquad\qquad (5.1)$$

If the physical constants are to be unity, then time, T, must be the same stuff as length, L, and mass, M, must be the inverse of length, L^{-1}.

$$[T] = [L] \qquad\qquad [M] = \frac{1}{[L]} \qquad\qquad (5.2)$$

Physically, M^{-1} is the Compton wavelength of a particle of mass M. As well as length, we need a mass dimension for electric charge, $[C]$.

It might be that physical constants come directly from mathematics. There are special numbers in mathematics. The number $\pi = 3.1415...$ is one such number; another such number, and a more important one, is $e = 2.71828....$ It might be that we will eventually have a "Theory of Everything" that starts with no more than the proposition that numbers exist and, from this, we can calculate the physical constants of nature as special numbers like π or e or $\sin(\sin(1))$.

Aside: In the novel "The Hitchhiker's Guide to the Galaxy', a computer is asked to produce such a theory of everything. It does as it is asked. The theory it produces is the number 42. At the frontiers of research into the nature of empty space (and everything else while we are at it), we are starting to think that the theory of everything might be the number $e = 2.71828$[1].

If the physical constants are from "special" numbers within mathematics, we are now no closer to understanding this or knowing where to find those numbers than we were 1,000,000 pints of water before.

It might be that the values of the physical constants are from initial conditions at the start of the universe. Perhaps they were set (by accident almost) when the big bang occurred 13.8 billion years ago and have remained unchanged since then. If this is so, it is unlikely that any theory we develop will enable us to calculate the values of the physical constants and the "theory of everything" will be a "theory of almost everything".

It might be that the values of the physical constants are from boundary conditions in the present day universe. It might be that the charge of the electron is 1 divided by the number of electrons in the

---

[1] The idea is that exponentiation of finite groups of permutation matrices generates all the different types of space, and the different types of space generate everything in the universe. The universe is no more than interacting types of empty space.

universe. It might be that the gravitational constant is determined for each galaxy group by the mass of that galaxy group – there's a Machian view.

The physical constants might not be constant. Perhaps they vary with the age of the universe, or perhaps they vary from place to place in the universe, or perhaps they vary with respect to some other parameter. All of these seem unlikely. The magnetic moment of the electron has been shown not to have varied by more than 1 part in a 1000 over the entire history of the universe; by modern standards, this is not very precise, but it is there. Observations of the Crab Nebula[2] pulsar show that the speed of light is constant (at visible frequencies) to within 5 parts in 10^{18}. The speed of light is also the same for different frequencies of electromagnetic radiation. Experiments at the two-mile linear accelerator at Stanford[3] in the USA show that visible light and gamma rays move at the same speed to within 6 parts in 10^6, and observations of radio waves and visible light from stellar flares[4] have shown that these two frequencies to travel at the same speed to within 1 part in 10^6. If the constants of nature were to change over time, then a stationary observer would see them change more slowly for a moving observer than for himself. After a while, the physics of the moving world would be different from the physics of the stationary world, and we would be in a right mess, unless the constants change in a way correlated to avoid such a mess, of course.

It might be that the physical constants are all tied together in some way whereby, if one varies, the others all vary to compensate, and any attempt to measure the variation of the one is cloaked by the compensatory variation of the others. In this scenario, the values of the physical constants can randomly vary in both space and time without this variation being detectable. If it is not detectable, is it meaningful?

The physical constants are numbers; they are not vectors or spinors or rank-two tensors; further, they are real numbers rather than complex numbers. Why are physical constants not complex

[2]. B. Warner & R.E. Nather (1969) Nature London, 222, 157.
[3]. B.C. Brown et al. Physical Review letters, 30, 763 (1973).
[4]. B. Lovel et al Nature, London, 202, 377 (1964).

numbers? – there's a question. As we will see later, the components of a vector are different when the vector is viewed in different co-ordinate systems. This does not happen with real numbers which are the same in all co-ordinate systems. Complex numbers cannot be physical constants because the components of the complex number change with a change in co-ordinate system, like a rotation – there's an answer. When dealing with changes of co-ordinate systems (and rotation in space or in space-time is a change of co-ordinate axes), single numbers are called scalars (as opposed to vectors or tensors). Physical constants have to be scalars (real numbers) if they are not to change under rotation of co-ordinate axes.

When we study 4-vectors, we will see that 4-acceleration in space-time is "perpendicular" to 4-velocity for stationary observers. Thus the velocity 4-vector is pulled to the side (rotated) by the 4-acceleration but not lengthened. Thus it is that magnitude of the 4-velocity through space-time cannot be changed (this is not velocity through space). The value of that 4-velocity through space-time is the speed of light. It is constant because the nature of space-time is such that 4-acceleration is "perpendicular" to 4-velocity. This explains why the speed of light constant, but thisit does not explain the particular value.

Aside: In 1913, Max Planck combined the constants of nature, $\{G,c,\hbar\}$ together to create natural units of length, time, and mass.

The Planck length is: $\degree = \left(\dfrac{\ }{\ }\right)^{-} = 2\times10^{35}$ Meters. The Planck

mass is: $m^\degree = \left(\dfrac{\hbar c}{G}\right)^{\frac{1}{2}} = 2\times10^{-8}$ Kilograms. The Planck time is:

$t^\degree = \dfrac{a^\degree}{c} = 10^{-43}$ Seconds.

CHAPTER 6

NUMBERS

What is a number? We have just considered the physical constants that have the same value when the axes of the co-ordinate system are changed (rotated perhaps), and we have said that these physical constants are (real) numbers, but what is a number?

The reader will be aware of at least two different types of numbers: the real numbers, $\mathbb{R}$, and the complex numbers, $\mathbb{C}$. What is it that these two different types of mathematical objects have in common so that we call them both numbers? What is it about blue stars and red stars that persuades us to call them both stars? Both blue and red stars have certain properties in common: they come out at night and they twinkle. So it is with numbers. As any mathematics student will tell you, a set of numbers is a division algebra. To a mathematics student, this means that the set (of numbers) satisfies thirteen of the fourteen axioms of an algebraic field.

It is outside of the subject matter of this book to discuss the details of the abstract algebra of algebraic fields. However, we give a flavour of what these axioms are like:

1. Multiplicative closure – a number multiplied by a number produces a number. It need not be this way. A vector dotted with a vector produces a number, not a vector.

2. Absence of zero-divisors – if the multiplicative product of two numbers is zero, then one, or both, of these numbers

must be zero. There are mathematical constructions that do not have this property; Clifford algebras are an example.

3. A most important algebraic field axiom is the one that insists that all the numbers, except zero, have a multiplicative inverse. The multiplicative inverse of 5 is $\frac{1}{5}$ because $5 \times \frac{1}{5} = 1$. A set of mathematical objects that have the property of each having a multiplicative inverse is called a division algebra (because we can do division inside it).

The fourteenth algebraic field axiom that need not be satisfied by a division algebra is multiplicative commutativity. Multiplicative commutativity is that two numbers produce the same number regardless of the order in which they are multiplied: $5 \times 3 = 3 \times 5$. Not all mathematical objects do this. Matrices are not generally commutative, but, more importantly for us, quaternions (see later) are not multiplicatively commutative. It is generally accepted that a division algebra is a type of numbers. So, in short, a type of numbers is a set of mathematical objects that satisfy the specific thirteen of the fourteen algebraic field axioms that form a division algebra. This is a precise way of saying we can do the mathematical operations of addition and multiplication within the division algebra without any problems.

A division algebra is almost the same thing as an algebraic field. The only difference is that algebraic fields are multiplicatively commutative whereas division algebras might or might not be multiplicatively commutative. Division algebras are often referred to as non-commutative algebraic fields.

Examples of division algebras are:

Commutative algebraic fields:

The real numbers, $\mathbb{R}$.

The complex numbers, $\mathbb{C}$.

The hyperbolic complex numbers, $\mathbb{S}$.

Non-commutative algebraic fields:

The quaternions, $\mathbb{H}$.

The anti-quaternions, $\mathbb{H}_{Anti}$.

The A_3 algebras.

Now, matrices in general satisfy only seven of the fourteen algebraic field axioms, but special sets of matrices satisfy either all the axioms of an algebraic field or all the axioms of an algebraic field except multiplicative commutativity. These special sets are each a division algebra and thus each set is a type of numbers. It is much easier to work with matrices than with other notations for the various types of numbers. We will therefore write our numbers as matrices. Examples are:

$$\mathbb{R} \equiv [a], \quad \mathbb{C} \equiv \begin{bmatrix} a & b \\ -b & a \end{bmatrix}, \quad \mathbb{S} \equiv \exp\left(\begin{bmatrix} a & b \\ b & a \end{bmatrix}\right) \tag{6.1}$$

$$\mathbb{H} \equiv \begin{bmatrix} a & b & c & d \\ -b & a & -d & c \\ -c & d & a & -b \\ -d & -c & b & a \end{bmatrix}$$

Which are respectively, the real numbers, the (Euclidean) complex numbers, the hyperbolic complex numbers, and the quaternions.

The reader will soon discover that we will be writing finite groups as sets of matrices. There is a very close connection between the finite groups and the types of numbers, and this connection is made patent by the use of matrix notation.

EXERCISES

1. Using the usual notation, first calculate $(a + ib)(c + id)$ and secondly calculate $\begin{bmatrix} a & b \\ -b & a \end{bmatrix}\begin{bmatrix} c & d \\ -d & c \end{bmatrix}$.

2. Is $\begin{bmatrix} 0 & 1 \\ 1 & 0 \end{bmatrix}$ a square root of plus unity?

3. Is the matrix form $\begin{bmatrix} a & b & 0 \\ 0 & a & b \\ b & 0 & a \end{bmatrix}$ multiplicatively closed - do two of them multiplied together make another of them?

4. What is the determinant of $\begin{bmatrix} 2 & 2 \\ 2 & 2 \end{bmatrix}$? Can we form a division algebra with these matrices?

5. Can the matrix $\exp\left(\begin{bmatrix} a & b \\ b & a \end{bmatrix}\right)$ ever have a zero determinant?

7

COMMENTS ON MATRICES

Imagine a co-ordinate system (a 2-dimensional Cartesian system is easiest) and a point, (x, y), in that co-ordinate system. If we wish to relate this point to another point, (a, b), in the co-ordinate system, then we must write, (x, y), in terms of (a, b). We do not have to make the relationship linear, but, if we do not make it linear, then we cannot use matrices to express the relationship, and so we will assume that the relationship is going to be linear. That is, we are going to assume that the relationship is of the form:

$$x = ka + lb$$
$$y = ma + nb$$

(7.1)

where $\{k, l, m, n\} \in \mathbb{R}$ are just real numbers. It seems that linearity is central to the nature of geometric space[1]. The reader might wish to try constructing a relationship between two points in space that cannot be expressed as linear relations to each other. If the reader does so attempt, she will understand why we think geometric spaces are basically linear.

Linearity is dealt with mathematically by the use of matrices. If the reader is unfamiliar with matrices, she is directed to the many texts upon this area of math. In the next paragraph, we skimp through matrix mathematics.

---

[1] Non-linear space, such as the curved space of general relativity, is a distortion of a linear space.

Matrices are "boxes of numbers'. Matrices are added component-wise, as you would expect them to be, but matrix multiplication
is not component-wise. Instead, matrix multiplication is linear multiplication (a row multiplies a column). Some examples:

$$\begin{bmatrix} a & b \end{bmatrix} + \begin{bmatrix} c & d \end{bmatrix} = \begin{bmatrix} a+c & b+d \end{bmatrix}$$

$$\begin{bmatrix} a & b \\ c & d \end{bmatrix} + \begin{bmatrix} e & f \\ g & h \end{bmatrix} = \begin{bmatrix} a+e & b+f \\ c+g & d+h \end{bmatrix} \tag{7.2}$$

$$\begin{bmatrix} a & b \end{bmatrix} \begin{bmatrix} c \\ d \end{bmatrix} = \begin{bmatrix} ac+bd \end{bmatrix}$$

$$\begin{bmatrix} a & b \\ c & d \end{bmatrix} \begin{bmatrix} e & f \\ g & h \end{bmatrix} = \begin{bmatrix} ae+bg & af+bh \\ ce+dg & cf+dh \end{bmatrix} \tag{7.3}$$

$$\begin{bmatrix} k & l \\ m & n \end{bmatrix} \begin{bmatrix} a \\ b \end{bmatrix} = \begin{bmatrix} ka+lb \\ ma+nb \end{bmatrix} = \begin{bmatrix} x \\ y \end{bmatrix}$$

Notice how, in the last example, we have "moved" the point (a, b)
to the point (x, y) mentioned above. We can think of matrices as
movements in space. (There are different types of space, of course.)
A matrix moves a point in space to another point in space. However,
for this to make sense, we need both the starting point and the ending point to be in the same type of space. In the last example, we
moved the point (a, b) in 2-dimensional space to the point (x, y) in
2-dimensional space, but, as we will see later, there is more than
one type of 2-dimensional space. (There are two different types of
2-dimensional space, $\{\mathbb{C}, \mathbb{S}\}$, in the finite groups.) How do we ensure
that the point is moved into the same type of space? The answer is
that the movement itself has to be part of the particular space. This
is a convoluted way of saying that the whole set of matrices, both
the points and the movement need to be in the same algebra – the
matrices need to be of the same form. Consider:

$$\begin{bmatrix} a & b \\ -b & a \end{bmatrix} \begin{bmatrix} c & d \\ -d & c \end{bmatrix} = \begin{bmatrix} ac-bd & ad+bc \\ -(ad+bc) & ac-bd \end{bmatrix} \tag{7.4}$$

All three matrices in this are of the same form. In fact, this is the
complex numbers, $\mathbb{C}$. The space associated with this matrix form is
the complex plane which is the 2-dimensional Euclidean space.

The different elements in a matrix are related to each other. In the example above, we have the relationship between the off-diagonal element and the diagonal element:

$$\begin{bmatrix} 0 & b \\ -b & 0 \end{bmatrix}\begin{bmatrix} 0 & b \\ -b & 0 \end{bmatrix} = \begin{bmatrix} -b^2 & 0 \\ 0 & -b^2 \end{bmatrix} \tag{7.5}$$

where the off-diagonal element is the square root of minus the diagonal element. This is more familiar as $i = \sqrt{-1}$.

It is possible to put complex numbers into matrices rather than real numbers, but this is just a notational shortcut. Unfortunately, it is a notational shortcut that might lead the student astray. We have:

$$\begin{bmatrix} a+ib & c+id \\ -c+id & a-ib \end{bmatrix} \equiv \begin{bmatrix} \begin{bmatrix} a & b \\ -b & a \end{bmatrix} & \begin{bmatrix} c & d \\ -d & c \end{bmatrix} \\ \begin{bmatrix} -c & d \\ -d & -c \end{bmatrix} & \begin{bmatrix} a & -b \\ b & a \end{bmatrix} \end{bmatrix}$$

$$= \begin{bmatrix} a & b & c & d \\ -b & a & -d & c \\ -c & d & a & -b \\ -d & -c & b & a \end{bmatrix} \tag{7.6}$$

We can always expand a matrix with complex elements into a matrix with real elements. The block multiplication properties of matrices guarantee this.

7.1 SYMMETRIC MATRICES AND ANTI-SYMMETRIC MATRICES

The transpose of a matrix is where the elements of the matrix have their line and column indices swapped, $\alpha_{rc} \rightarrow \alpha_{cr}$, for example:

$$\begin{bmatrix} a & b & c \\ d & e & f \\ g & h & j \end{bmatrix}^{Transpose} = \begin{bmatrix} a & d & g \\ b & e & h \\ c & f & j \end{bmatrix} \tag{7.7}$$

Transposing matrices is not a proper algebraic operation; it is a notational shortcut. It works, "by accident" when we are working in n-dimensional spaces that can be divided into $(n-1)$ 2-dimensional sub-spaces by combining any imaginary axis with the real axis. It also works in $\mathbb{R}^n$ spaces. The reader will see it often. Since we will be concerned with only the spaces that derive from the finite group C_2 and its cross products, and since these are spaces that can be divided into $(n-1)$ 2-dimensional sub-spaces by combining any imaginary axis with the real axis, it will work for us. It allows us to deal with the 2-dimensional form of symmetry easily. Symmetry and anti-symmetry (the 2-dimensional form) will be important later in this book when we consider 4-dimensional space-time and the Lorentz group. Be warned; matrix transposition does not work generally in spaces of more than two dimensions.

Aside: The finite group C_3 contains 3-dimensional symmetries which are very different from the 2-dimensional concept of symmetry to which we are accustomed. The same is true of all the other finite groups that are not products of C_2. Symmetry and anti-symmetry mean different things in different types of space.

Aside: A matrix is called unitary if the product of the matrix with its transpose is equal to the identity matrix (the matrix with 1s on the leading diagonal and zeros everywhere els. E.g.:

$$\begin{bmatrix} 0 & 1 \\ -1 & 0 \end{bmatrix}\begin{bmatrix} 0 & -1 \\ 1 & 0 \end{bmatrix} = \begin{bmatrix} 1 & 0 \\ 0 & 1 \end{bmatrix} \tag{7.8}$$

Unitary matrices play a central role in quantum field theory, and the reader is sure to meet them in later studies.

We say that a matrix is symmetric (again we mean 2-dimensionally anti-symmetric) if it is equal to its transpose. For example:

$$Symmetric \rightarrow \begin{bmatrix} a & b & c & d \\ b & a & d & c \\ c & d & a & b \\ d & c & b & a \end{bmatrix} \tag{7.9}$$

We say that a matrix is anti-symmetric (again we mean 2-dimensionally anti-symmetric) if it is equal to minus its transpose. For example:

$$Anti-symmetric \rightarrow \begin{bmatrix} 0 & b & c & d \\ -b & 0 & -d & c \\ -c & d & 0 & -b \\ -d & -c & b & 0 \end{bmatrix} \tag{6.10}$$

(This anti-symmetric matrix is a quaternion with zero real part.),

---

Aside: Symmetry and anti-symmetry are a very important part of our present understanding of the univeese. Electromagnetism is an anti-symmtrices force. The Maxwell equations of electromagnetism are later derived from the quaternions which are an anti-symmetric algebra based on anti-symmetric variables. It seems to be that because electromagnetism is an anti-symmetric force, we have electromagnetic forces that are both attractive and repulsive. General relativity uses tensors written as symmetric matrices. Gravity is a symmetric force, and this, it seems, is why gravitational force is attractive only. Within particle physics, much of the current understanding, for example the Higgs mechanism that gives mass to particles, is based upon the idea of symmetry being broken. Within cosmology, the phase transitions postulated to have happened at the birth of the universe are similarly associated with symmetry breaking.

7.2 ROTATION MATRICES

We form rotation matrices by taking the exponential of the non-real variable(s) of an algebra. If these variables are 2-dimensionally anti-symmetric, we get a Euclidean type rotation matrix containing Euclidean trigonometric functions and we refer to this as an anti-symmetric rotation. If these variables are symmetric, we get a hyperbolic type (space-time) rotation matrix containing hyperbolic trigonometric functions and we refer to this as a symmetric rotation. Rotation in space is done with a rotation matrix that is anti-symmetric in the angle. Rotation in space-time is done with a rotation matrix that is symmetric in the angle.

It is possible to get a rotation matrix in higher dimensional spaces that is formed by exponentiating both anti-symmetric variables and symmetric variables. Such a rotation is then a mixed symmetric and anti-symmetric rotation. We will meet such algebras later (the A_3 algebras). In such a rotation matrix, we get that the commutator of two consecutive different symmetric rotations are equivalent to an anti-symmetric rotation. In the conventional parlance, "two space-time boosts make a space rotation'. This cannot happen in 2-dimensional space where two space-time boosts make a third space-time boost.

We end this chapter with a couple of asides that the reader does not need to know in order to understand special relativity. They are included because they are interesting. There is more to life than special relativity.

Aside: Unitary[2] matrices are usually written with complex elements. Symplectic[3] matrices are usually written with quaternion elements. Both types of matrix can be written using real elements, and doing so simplifies everything in the universe.

Aside: By choosing the appropriate basis, a unitary matrix can be written with zeros everywhere except on the leading diagonal. For example:

$$\begin{bmatrix} a+ib & 0 & 0 & 0 \\ 0 & c+id & 0 & 0 \\ 0 & 0 & e+if & 0 \\ 0 & 0 & 0 & g+ih \end{bmatrix} \tag{7.11}$$

This matrix is a movement in $\mathbb{C}^4$ space. Symplectic matrices have the "same" property but with quaternions on the leading diagonal instead of complex numbers.

[2.] A unitary matrix is a matrix set in $\mathbb{C}^n$ space.
[3.] A symplectic matrix is a matrix set in $\mathbb{H}^n$ space.

EXERCISES

1. Is the rotation matrix $\begin{bmatrix} \cosh\chi & \sinh\chi \\ \sinh\chi & \cosh\chi \end{bmatrix}$ an anti-symmetric matrix?

2. Is the rotation matrix $\begin{bmatrix} \cos\theta & \sin\theta \\ -\sin\theta & \cos\theta \end{bmatrix}$ an anti-symmetric matrix? What about when $\theta = \dfrac{\pi}{2}$?

3. Is the rotation matrix $\begin{bmatrix} \cos\theta & \sin\theta \\ -\sin\theta & \cos\theta \end{bmatrix}$ an anti-symmetric in the variable θ?

CHAPTER 8

INTRODUCTION TO FINITE GROUPS

Finite groups, together with the real numbers and the exponential function (and arguably the concept of linearity), are the basis of mathematics. Except for areas of mathematics invented by humankind, all of algebra, geometry, calculus, and much other stuff is based on no more than these three things. Like numbers, and unlike 3-dimensional Riemannian space or Clifford algebras[1], in the same way that numbers really exist, finite groups really exist. We find them lying about in the universe, as we do with real numbers and the exponential function. Since mathematics is essential to physics, finite groups, the real numbers, and the exponential function, are of the essence of physics.

Since finite groups really exist, the geometric spaces within them, the finite group spaces, really exist, and any theory of the universe must include an understanding of these spaces. It turns out that the finite group spaces derived from the cyclic group C_2 and its cross-product groups like $C_2 \times C_2$ are the spaces of the observable universe. The space-time of special relativity is no more than a space from the finite group C_2, and special relativity derives directly from this group (we have to add in the concepts of mass and of electric

---

[1] Actually Clifford algebras really exist but not as they are usually presented. See: Dennis Morris: The Naked Spinor – a Rewrite of Clifford algebra ISBN: 978-1507817995.

charge). The group C_2 is no more than the numbers $\{+1, -1\}$; it is hard to think of anything more simple and amazing that space and time are within only these two numbers. So, let us proceed to build the universe from no more than C_2.

Aside: Finite groups were discovered by Evariste Galois. Galois was born at Bourg-la-Reine near Paris on 25th October 1811. For the first twelve years of his life, he was educated by his mother who had a considerable knowledge of both religion and the classics. He entered the Lycee Louis-le-Grand in 1823 where he immediately found himself in the center of rebelling students refusing to chant in chapel; a hundred of them, not including Galois, were expelled. In this period, it is said of him that he read Legendre's "Elements de Geometrie" like a novel and mastered it in one reading; it is a degree level text and equivalent to three years of normal university study. At the age of fifteen, he was reading mathematical research papers. Although Galois failed the entrance examination for the Ecole Poly-techique, Terquem (editor of Nouvelles Annals des Mathematiques) puts it down to "... A candidate of superior intelligence is lost with an examiner of inferior intelligence...". His second attempt to gain entrance to the Ecole Polytechnique failed when he upset the examiner, Dinet, by correcting him on his understanding of the nature of logarithms.

8.1 HOW TO FIND FINITE GROUPS

Take square matrices and put 1s into them in such a way that there is only a single 1 in each row and only a single 1 in each column. There is only one such 1×1 matrix; it is just $[1]$; this is the group C_1. There are two, and only two, such 2×2 matrices:

$$\begin{bmatrix} 1 & 0 \\ 0 & 1 \end{bmatrix}, \begin{bmatrix} 0 & 1 \\ 1 & 0 \end{bmatrix} \tag{8.1}$$

These are the group C_2. With 3×3 matrices, we get:

$$\begin{bmatrix} 1 & 0 & 0 \\ 0 & 1 & 0 \\ 0 & 0 & 1 \end{bmatrix}, \begin{bmatrix} 0 & 1 & 0 \\ 0 & 0 & 1 \\ 1 & 0 & 0 \end{bmatrix}, \begin{bmatrix} 0 & 0 & 1 \\ 1 & 0 & 0 \\ 0 & 1 & 0 \end{bmatrix}$$
$$\begin{bmatrix} 1 & 0 & 0 \\ 0 & 0 & 1 \\ 0 & 1 & 0 \end{bmatrix}, \begin{bmatrix} 0 & 1 & 0 \\ 1 & 0 & 0 \\ 0 & 0 & 1 \end{bmatrix}, \begin{bmatrix} 0 & 0 & 1 \\ 0 & 1 & 0 \\ 1 & 0 & 0 \end{bmatrix}$$

(8.2)

There are six such matrices. Multiplying these sets of matrices together, we discover that they form a multiplicatively closed sets – that is they all multiply into one-another. Each such set also contains the multiplicative inverses (a matrix multiplied by its multiplicative inverse produces the multiplicative identity matrix) of every element of the set, and the set contains the multiplicative identity (the equivalent of the number +1). We demonstrate:

$$\begin{bmatrix} 1 & 0 \\ 0 & 1 \end{bmatrix} \begin{bmatrix} 0 & 1 \\ 1 & 0 \end{bmatrix} = \begin{bmatrix} 0 & 1 \\ 1 & 0 \end{bmatrix}$$

(8.3)

$$\begin{bmatrix} 1 & 0 \\ 0 & 1 \end{bmatrix}^2 = \begin{bmatrix} 1 & 0 \\ 0 & 1 \end{bmatrix}, \begin{bmatrix} 0 & 1 \\ 1 & 0 \end{bmatrix}^2 = \begin{bmatrix} 1 & 0 \\ 0 & 1 \end{bmatrix}$$

$$\begin{bmatrix} 0 & 1 & 0 \\ 0 & 0 & 1 \\ 1 & 0 & 0 \end{bmatrix}^2 = \begin{bmatrix} 0 & 0 & 1 \\ 1 & 0 & 0 \\ 0 & 1 & 0 \end{bmatrix}, \begin{bmatrix} 0 & 0 & 1 \\ 1 & 0 & 0 \\ 0 & 1 & 0 \end{bmatrix}^3 = \begin{bmatrix} 1 & 0 & 0 \\ 0 & 1 & 0 \\ 0 & 0 & 1 \end{bmatrix}, \ \dots \ (8.4)$$

The reader might want to complete the set of 3×3 multiplications. These $n \times n$ matrices with a single 1 on every row and a single 1 in every column are called permutation matrices. They are so named because they correspond one-to-one with the possible ways of permuting n objects.

The important point is that each set of permutation matrices is a multiplicatively closed set that includes multiplicative inverses and the identity. (The identity is the matrix with all the 1s on the leading diagonal). Each of the above individual sets of (multiplicatively closed) $n \times n$ permutation matrices a finite group. Finite group are common called just groups[2].

Although we represent a group with permutation matrices, a group is really a set of (multiplicative) relations that are (multiplicatively) closed. I'll say that again: A group is a particular set of relations between objects. For the group of 2×2 matrices above, these relations are $\{a^2 = a, b^2 = a, ab = ba = b\}$. These relations are exactly the same multiplicative relations as the multiplicative relations between the numbers $\{+1, -1\}$ which are $\{(+1)^2 = +1, (-1)^2 = +1, (+1)(-1) = (-1)(+1) = -1\}$. It is the differences in the multiplicative relations between the objects in a group that distinguishes one group from another and not the nature of the objects that are being multiplied together.

Mathematicians refer to the objects (matrices) in a group as being a group when they really mean that the relations between the objects are the group. This happens because the objects bring with them the relations between them. A complete set of $n \times n$ permutation matrices represents a group, and a complete set of permutations of n objects also represents the same group because they have the same relations between them. It is because there is a finite number of objects (permutations or matrices) in the group, that they are called finite groups. Having rattled on about relations between things, your author admits that he thinks of groups as sets of permutation matrices because it is easier than thinking of them as sets of abstract relations. No-harm will come from the reader doing the same.

It is also possible to think of the cyclic group C_2 as the two square roots of plus-one and to think of the cyclic group C_3 as the three cube roots of plus-one and the cyclic group C_4 as the four fourth roots of plus-one and so on for all the cyclic groups.

The order of the group is the number of objects (matrices) in the group. There is at least one group of every order, and there is often more than one group of a given order. There are different types of groups – different structures of the multiplicative relations between the permutation matrices. If the group order is a prime number,

---

[2.] There are infinite groups that contain an infinite number of objects. The unit circle in the complex plane is one. It is known as $U(1)$.

there is only one group of that order and that is of a type of group known as a cyclic group.

8.2 SUB-GROUPS AND TYPES OF ROTATION

Within the six 3×3 permutation matrices above, there are three matrices which are a multiplicatively closed set on their own:

$$\begin{bmatrix} 1 & 0 & 0 \\ 0 & 1 & 0 \\ 0 & 0 & 1 \end{bmatrix}, \begin{bmatrix} 0 & 1 & 0 \\ 0 & 0 & 1 \\ 1 & 0 & 0 \end{bmatrix}, \begin{bmatrix} 0 & 0 & 1 \\ 1 & 0 & 0 \\ 0 & 1 & 0 \end{bmatrix} \tag{8.5}$$

These too are a group. As well as being a group in their own right, these three permutation matrices are also a sub-group of the six 3×3 permutation matrices. The multiplicative relations of this order 3 group are:

$$\{a^2 = a^3 = a, b^3 = c^3 = a, ab = b, ac = c,$$
$$bc = cb = a, b^2 = c, c^2 = b\} \tag{8.6}$$

The sub-groups of a group correspond to sub-spaces of the geometric space from the whole group. For example, the order four group $C_2 \times C_2$ contains three order two sub-groups. Because of this, the 4-dimensional spaces from this group contain three 2-dimensional sub-spaces. This means that it is possible to perform 2-dimensional rotations in three different planes within the spaces from this group. It is not possible to perform a 2-dimensional rotation in any 3-dimensional space because the only order three group, C_3, has no order two sub-group. We can 2-dimensionally rotate in three different planes within the space in which we sit because the space in which we sit is from the group $C_2 \times C_2$. If it were that the space in which we sit was from the group C_4, we would be able to perform 2-dimensional rotation in only one of the three possible planes because C_4 has only one order two sub-group.

8.3   GEOMETRIC SPACES FROM FINITE GROUPS

Now, since groups can be represented by multiplicatively closed sets of matrices, they can also be thought of as multiplicatively closed sets of linear transformations. Linear transformations are movements in geometric space. Since these movements are closed (matrix multiplication) and, it turns out, independent of each other basis of a geometric space. We will see the details later.

8.4   MORE ON GROUPS

Groups come in families, and each finite group has a name. The set of 2×2 permutation matrices is called the cyclic group of order 2 and is denoted as C_2. This is the group of central interest to us. The set of six 3×3 permutation matrices is called the symmetric group of order 3 and is denoted as S_3. The multiplicatively closed set of three 3×3 permutation matrices shown above is called the cyclic group of order 3 and is denoted as C_3. The cyclic group of order 2 is sometimes also known as the symmetric group of order 2 and denoted as S_2. Note that the symmetric groups, S_n, are the complete sets of permutations of n objects. There are $n!$ such permutations for each n and so the group S_n is of order $n!$. For example: S_3 is the set of permutations of three objects but it has order six. Another family of groups is the alternating groups, A_n, which are the sets of even permutations and thus of order $\dfrac{n!}{2}$.

8.5   NON-COMMUTATIVITY AND THE SPACE IN WHICH WE SIT

Being like matrices, some groups are multiplicatively commutative - $ab = ba$ - and some groups are not multiplicatively commutative – $ab \neq ba$. We call the commutative groups abelian groups and we call the non-commutative groups non-abelian groups. The 4-dimensional space in which we sit is a non-commutative space as is the quaternion space that holds electromagnetism. It is remarkable that these non-commutative spaces come out of the commutative group $C_2 \times C_2$.

The First Finite Groups

Abelian Groups	Non-A Groups	Abelian Groups	Non-A Groups
C_1		$C_9, C_3 \times C_3$	
$C_2, \cong S_2$		C_{10}	D_5
C_3		C_{11}	
$C_4, C_2 \times C_2$		$C_{12}, C_2 \times C_6$	$A_4, D_6 \times T$
C_5		C_{13}	
C_6	$D_3 \cong S_3$	C_{14}	D_7
C_7		C_{15}	
$C_8, C_2 \times C_4,$ $C_2 \times C_2 \times C_2$	D_3, Q		

There are too many groups of order 16 to fit in the table.

8.6 THE FINITE GROUP MATRIX

The group of six 3×3 matrices, S_3, can also be written as six 6×6 matrices – it is of order 6. If we do this, we can fit all the six elements of the group into one matrix where we represent each permutation matrix by a different letter:

$$
\begin{bmatrix}
a & b & c & d & e & f \\
c & a & b & e & f & d \\
b & c & a & f & d & e \\
d & e & f & a & b & c \\
e & f & d & c & a & b \\
f & d & e & b & c & a
\end{bmatrix}
\tag{8.7}
$$

Notice that all the *a* s are on the leading diagonal. This can be done with any finite group[3]. Other examples are:

$$C_2 = \begin{bmatrix} a & b \\ b & a \end{bmatrix}, \quad C_3 = \begin{bmatrix} a & b & c \\ c & a & b \\ b & c & a \end{bmatrix} \tag{8.8}$$

Thus, any finite group of order n can be written as a $n \times n$ matrix. We call such a matrix the finite group matrix. Furthermore, any such matrix, with each particular element appearing once and only once in each row and in each column and with all the *a* s on the leading diagonal, is a way of representing some group. It is by exponentiating the finite group matrix that we get a geometric space from that group. There's a bit more to it because a finite group contains many geometric spaces.

If we try to do this with 4×4 matrices, we find there are four such matrices. However, there are not four different order four groups here. There are just two different order four groups that are known as $C_2 \times C_2$ and C_4. One of the 4×4 matrices is the $C_2 \times C_2$ group, which is of interest to us. The other three 4×4 matrices are different ways of writing the group C_4, which is not of interest to us. Different groups are different because they have different multiplicative relations between their elements. In the $C_2 \times C_2$ group, the three $\{b, c, d\}$ elements all square to the a element.

In the C_4 group, only one of the $\{b, c, d\}$ elements squares to the a element.

There is only one group with five elements (it is C_5), but there are many ways of writing this group as a 5×5 matrix.

8.7   SPECIAL RELATIVITY IS IN C_2

The reader has now earned a little taste of what is to come. This is out of sequence within this book, and it will be explained in more detail later in the book, but here goes. Start with the group C_2 written

---

[3]. This is an easy way of constructing the Cayley table (multiplication table) of the group. We do not need to know about Cayley tables in this book.

as one matrix; consider each element to be a real number; exponentiate the matrix; and *voila*:

$$\exp\left(\begin{bmatrix} a & b \\ b & a \end{bmatrix}\right) = \begin{bmatrix} h & 0 \\ 0 & h \end{bmatrix}\begin{bmatrix} \cosh b & \sinh b \\ \sinh b & \cosh b \end{bmatrix} \tag{7.9}$$

This is the space-time (2-dimensional) of special relativity. Within this is the phenomena of time dilation, length-contraction etc. The determinant of the matrix gives the distance function of space-time. We will shortly have the special theory of relativity from no-more than the group C_2 and real numbers! (We have to add in mass and electric charge.) Further, we can think of C_2 as no more than the numbers $\{+1, -1\}$.

The element of the group on the leading diagonal (the a element) is called the identity element of the group because when $a = 1$ the product of it and any other element is that other element. It is the same thing to a group as the number 1^4 is to the real numbers. This becomes (after exponentiation) the radial co-ordinate of the (in polar co-ordinates) space.

Within the different algebras we will meet, we sometimes have elements that square to the identity, plus-one, and we sometimes have elements that square to minus the identity, minus-one. The complex numbers, $\mathbb{C}$, have the element $i^2 = -1$. The hyperbolic complex numbers have the element $r^2 = +1$. The quaternions have the elements $\{i, j, k\}$ which all square to minus-one. The A_3 algebra that we will meet later has two imaginary elements that square to plus-one and one imaginary element that squares to minus-one. Such relations are an expression of the 2-dimension symmetry (the elements that square to plus-one) and the 2-dimensional anti-symmetry (the elements that square to minus-one) of the algebra. Thus, looked at this way, (look at the matrices), the (Euclidean) complex numbers are an anti-symmetric 2-dimensional algebra, and the hyperbolic complex numbers are a symmetric 2-dimensional algebra. The anti-symmetric variables (those that square to minus-one) become trigonometric functions within a spatial rotation matrix. The symmetric variables (those that square to plus-one) become trigonometric functions within a space-time rotation matrix.

[4.] The number 1 is the multiplicative identity of the real numbers (and the complex numbers ....)

There is much more to finite group theory than mentioned above, but there is neither space for it, or need of it, in this book. However, I cannot speak of group theory without mention of the classification theorem. This is not part of what we need to understand special relativity, but but and then add knowing about it is a useful fact for the reader.

8.8　THE CLASSIFICATION OF THE FINITE GROUPS

Within the real numbers, there are prime numbers (like: 2, 3, 5, 7, 11, 13, 17 …) and composite numbers that are products of the prime numbers. Within the finite groups, there are "prime" groups called simple finite groups. The simple finite groups are the same to groups as prime numbers are to real numbers, but there is more than one type of simple finite group whereas there is only one type of prime number. There are three infinite families of simple finite groups and 26 one-offs.

Aside:
The Classification of Finite Simple Groups theorem

The simple finite groups are:

Infinite family:	The cyclic groups of prime order
Infinite family:	The alternating groups A_3 and all higher orders[5]
Infinite family:	The Chevalley, Twisted Chevalley, and the Tits group
Sporadics:	26 individual groups

The largest of the sporadic groups is called "the monster"; it is something to do with string theory – see the "monster moonshine theorem". The cataloguing of the finite simple groups took over 150 years. It is seen as the crowning achievement of 20th century mathematics.

---

[5]. A_5 is of order 60.

Since finite groups underlie geometric spaces, we now have a classification of all the possible types of geometric space, and, thus, of all possible universes!

Aside: The Odd Order Theorem, also known as the Feit-Thompson Theorem, says that:

except for the prime order cyclic groups, the order of every simple finite group must be even.

This theorem is an extremely important theorem in finite group theory. It took the efforts of many people over many decades to prove it. None-the-less, Feit and Thompson had to spend several years trying to find a journal in which to publish the proof because the proof was so long (255 pages). They eventually published in the little known Pacific Journal of Mathematics wherein the proof took a whole issue. So, the reader ought not to expect to find it easy to publish just because she produces the biggest breakthrough in human understanding that the world has ever seen.

EXERCISES

1. Calculate: $\exp\left(\begin{bmatrix} 0 & b \\ -b & 0 \end{bmatrix}\right)$.

2. Calculate: $\exp\left(\begin{bmatrix} 0 & b & 0 \\ 0 & 0 & b \\ b & 0 & 0 \end{bmatrix}\right)$.

CHAPTER 9

TRIGONOMETRIC FUNCTIONS

We are concerned with the geometric spaces that arise from the finite groups. We are not concerned with the geometric spaces of the form $\mathbb{R}^n$ made by fitting copies of $\mathbb{R}$ together. Although there are similarities between the spaces that arise from the finite group C_2 and the spaces based on $\mathbb{R}^2$, they are not the same thing. There are no such similarities in 3-dimensions, but there are some similarities in some cases in 4-dimensions.

Consider a circle of unit radius in the Euclidean plane and consider a point on that circle:

The Cosine and Sine Functions

The cosine of the angle is the horizontal distance from the vertical axis (of a unit circle) to the point on the circle. The sine of the angle is the distance in the vertical direction from the horizontal axis to the point on the unit circle. Thus, the unit circle is the set of points parameterized by $(x, y) = (\cos\theta, \sin\theta)$. Clearly,

$$\cos^2\theta + \sin^2\theta = x^2 + y^2 = 1 \qquad (9.1)$$

The sine and cosine were first defined this way in 6th century India.

The sine and cosine are the trigonometric functions of 2-dimensional Euclidean space. We will see later that they are associated with the complex plane, $\mathbb{C}$. In this work, we will meet other types of trigonometric functions that are associated with other types of space (space-time for example). Each of these trigonometric functions will be a projection on to an axis of that space. The trigonometric functions of 2-dimensional space-time are the cosh( ) and sinh( ) functions. The cosh( ) function is a projection on to the time axis of space-time, and the sinh( ) function is a projection on to the space axis of space-time. Clearly, 3-dimensional spaces, which have three axes, have sets of three trigonometric functions and 4-dimensional spaces, which have four axes, have sets of four trigonometric functions - these spaces are not the $\{\mathbb{R}^3, \mathbb{R}^4\}$ spaces. 1-dimensional space has one trigonometric function.

9.1   CIRCLES DEFINED

It is normal to define a unit circle as the locus of a point that is distance 1 from the origin. Obscurely, there is another, but equivalent, way of defining a unit circle. It is the locus of a point whose x co-ordinate is the rate of change of its y co-ordinate with respect to the angle made with the horizontal axis and whose y co-ordinate is the rate of change of its x co-ordinate with respect to the angle made with the horizontal axis. This says that the projections from a point on a unit circle, the trigonometric functions, must differentiate into each other (or perhaps the minus of each other), like $\dfrac{d}{d\theta}\sin\theta = \cos\theta$.

Thinking of a circle in this way, with a lot of thought, the reader will perhaps understand why the trigonometric functions are related to the exponential function, which differentiates into itself.

Trigonometric functions are based on "splittings" of the series of the exponential function. This is seen most clearly in the series expansions of the 2-dimensional hyperbolic trigonometric functions. We have:

$$\exp(x) = 1 + \frac{x}{1!} + \frac{x^2}{2!} + \frac{x^3}{3!} + \frac{x^4}{4!} + \frac{x^5}{5!} + \frac{x^6}{6!} + \frac{x^7}{7!} + \frac{x^8}{8!} + \frac{x^9}{9!} + \dots \quad (9.2)$$

$$\cosh(x) = 1 + \frac{x^2}{2!} + \frac{x^4}{4!} + \frac{x^6}{6!} + \frac{x^8}{8!} + \dots$$

$$\sinh(x) = \frac{x}{1!} + \frac{x^3}{3!} + \frac{x^5}{5!} + \frac{x^7}{7!} + \frac{x^9}{9!} + \dots$$

The sine and cosine functions are the same but with some minus signs thrown in:

$$\cos(x) = 1 - \frac{x^2}{2!} + \frac{x^4}{4!} - \frac{x^6}{6!} + \frac{x^8}{8!} - \dots \quad (9.3)$$

$$\sin(x) = \frac{x}{1!} - \frac{x^3}{3!} + \frac{x^5}{5!} - \frac{x^7}{7!} + \frac{x^9}{9!} - \dots$$

Now go back and re-read the paragraph above beginning "It is normal…". It is important to understand the relationship between the exponential function and circles (in different types of space).

When we define a circle in the obscure way, we are defining the trigonometric functions to be splittings of the series of the exponential function because only by splitting the exponential function series can we get functions that differentiate into each other. Those splittings of the exponential function series are the co-ordinates of the point on the circle and so are the projections from that point on to the axes of the space. Starting with the exponential function, we get trigonometric functions, which give us circles; this is another way of saying that we get rotation from the exponential function. Rotation brings with it space. Special relativity is about rotation in space-time, and so special relativity is about the exponential function – there's a thought!

There are only two ways of splitting the exponential function series into two functions that differentiate into each other. One way is the straight-forward splitting of the hyperbolic trigonometric functions, {cosh( ), sinh( )}. The other way is to throw a minus sign in before every second term of the two-way splittings of the exponential series and thereby produce the Euclidean trigonometric functions {cos( ), sin( )}.

Aside: If we were to split the exponential series into three, we would get three functions that differentiated into each other. If we were to throw in a minus sign before every second term in the three-way splittings of the exponential series, we would again get three functions that differentiate into each other; this is the basis of the 3-dimensional trigonometric functions of the C_3 geometric spaces.

Aside: In 1-dimensional space, the angle, x, is zero, and the 1-dimensional trigonometric function is 1, since $\exp(0) = 1$, which is obviously a projection on to the axis from the unit circle of 1-dimensional space.

Aside: In 3-dimensional space (derived from the finite group C_3), the trigonometric functions are matrix products of 3-way splittings of the exponential:

$$
\begin{bmatrix}
1+\dfrac{x^3}{3!}+\dfrac{x^6}{6!}+\dots & \dfrac{x}{1!}+\dots & \dfrac{x^2}{2!}+\dots \\[2ex]
\dfrac{x^2}{2!}+\dfrac{x^5}{5!}+\dfrac{x^8}{8!}+\dots & 1+\dots & \dfrac{x}{1!}+\dots \\[2ex]
\dfrac{x}{1!}+\dfrac{x^4}{4!}+\dfrac{x^7}{7!}+\dots & \dfrac{x^2}{2!}+\dots & 1+\dots
\end{bmatrix}
\tag{9.4}
$$

$$
\begin{bmatrix}
1+\dfrac{y^3}{3!}+\dfrac{y^6}{6!}+\dots & \dfrac{y^2}{2!}+\dots & \dfrac{y}{1!}+\dots \\[2ex]
\dfrac{y}{1!}+\dfrac{y^4}{4!}+\dfrac{y^7}{7!}+\dots & 1+\dots & \dfrac{y^2}{2!}+\dots \\[2ex]
\dfrac{y^2}{2!}+\dfrac{y^5}{5!}+\dfrac{y^8}{8!}+\dots & \dfrac{y}{1!}+\dots & 1+\dots
\end{bmatrix}
$$

The exponential "splittings" are more complicated for non-commutative groups.

9.2   ROTATION MATRICES

Every type of geometric space has associated with it a matrix (linear transformation) that "moves" a point (it moves the point by multiplying the matrix representing the point) in that space in such a way that the point is the same distance from the origin after the movement as the point was before that movement. Such a movement (linear transformation) is called a rotation. So, a rotation is just a movement that preserves the distance from the origin. The particular matrix that does the "rotational movement" is called a rotation matrix.

The rotation matrix of a space is calculated by exponentiating the finite group matrix that underlies the space. We demonstrate with the finite group C_2.

$$\exp\left(\begin{bmatrix} a & b \\ b & a \end{bmatrix}\right) = \begin{bmatrix} e^a & 0 \\ 0 & e^a \end{bmatrix} \begin{bmatrix} 1 + \dfrac{b^2}{2!} + \dfrac{b^4}{4!} + \dots & b + \dfrac{b^3}{3!} + \dfrac{b^5}{5!} + \dots \\ b + \dfrac{b^3}{3!} + \dfrac{b^5}{5!} + \dots & 1 + \dfrac{b^2}{2!} + \dfrac{b^4}{4!} + \dots \end{bmatrix} \tag{9.5}$$

$$= \begin{bmatrix} h & 0 \\ 0 & h \end{bmatrix} \begin{bmatrix} \cosh b & \sinh b \\ \sinh b & \cosh b \end{bmatrix}$$

The matrix containing the $\{\cosh b, \sinh b\}$ functions is the rotation matrix of 2-dimensional space-time. As well as a matrix that rotates a point, we have a matrix $\{h\}$ that moves the point toward or away from the origin, and so every point in space-time can be denoted by values of $\{h, b\}$. This polar form is simply space-time (hyperbolic complex numbers).

Aside: The hyperbolic complex numbers, $\mathbb{S}$, are a handed algebraic field. A handed algebraic field is an algebraic field which has no additive inverses on the real axis. They are the same thing as space-time in the same way that the Euclidean complex numbers, $\mathbb{C}$, are the same thing as the Euclidean plane. Note that the Cartesian form of the hyperbolic complex numbers is not a handed algebraic field (it contains singular matrices); only the polar form is a handed algebraic field. We could be rid of the handedness by putting a ± in front of the polar form, but the mathematics does not give us this, and so we go with the math. We cannot travel backwards in time, and so the math is right.

Now, each hyperbolic angle, χ, corresponds to a velocity through space-time. The hyperbolic complex numbers (space-time numbers) do not include the matrix $\begin{bmatrix} 1 & 1 \\ 1 & 1 \end{bmatrix}$ as this would mean that cosh $\chi = \sinh \chi$ for some hyperbolic angle, χ, which cannot be the case. This means that the hyperbolic angle that corresponds to the velocity of light is not part of the hyperbolic complex number algebra. Thus, it seems, light is outside of space-time.

Aside: The hyperbolic complex numbers, $\mathbb{S}$, were first discovered by Cockle[1] (1819–1895) in 1848 (who was later to become the president of the London Mathematical Society 1886–1888). They have since been independently discovered at least thirfifteen times and have been given at lst thirteenen different names. It is remarkable that, although the Euclidean complex numbers, $\mathbb{C}$, are widely taught, the hyperbolic complex numbers are taught hardly anywhere. The various names are:

Tessarines	Cockle[2]	1848
Algebraic motors	Clifford[3]	1882
Hyperbolic complex numbers	Vignaux[4]	1935
	Sobczyk[5]	1995
	Guo Chun Wen	2002

---

[1] On certain functions resembling quaternions and on new imaginary algebra. Phil. Mag. (3), 33, pp. 435–439.
And On a new imaginary in algebra Phil. Mag. (3), 34, pp. 37–47.
[2] On the symbols of algebra, and on the theory of tessarines Phil. Mag. (3), 34, pp. 406–410
[3] Clifford, W.K. Mathematical Works (1882) pp. 392 "Further notes on biquaternions".
[4] Sobre el numero complejo hiperbolico y su relacion con la geometria de Borel. Universidad Nacional de la plata Republica Argentina
[5] Sobczyk, G (1995) Hyperbolic number plane College Mathematics journal 26:268–80.

Bireal numbers	Bencivenga[6]	1946
Approximate numbers	Warmus[7]	1956
Double mbers	Yaglornm[8]	1968
	Hazewinkle[9]	1990
Anormal complex numbers	Benz[10]	1973
Perplex numbers	Fjelstad[11]	1986
Lorentz numbers	Harvey	1990
Dual numbers	Hucks	1993
Split-complex numbers	Rosenfield[12]	1997
Study numbers	Lounesto[13]	2001
Twocomplex numbers	Olariu[14]	2002

It is a fact that the complex numbers, $\mathbb{C}$, are most easily written as a 2×2 matrix:

$$a + ib \cong \begin{bmatrix} a & b \\ -b & a \end{bmatrix} \tag{9.6}$$

This is the origin of the "weird" multiplication operation of the complex numbers. Note:

$$\begin{bmatrix} a & b \\ -b & a \end{bmatrix}\begin{bmatrix} c & d \\ -d & c \end{bmatrix} = \begin{bmatrix} ac - bd & ad + bc \\ ad + bc & ac - bd \end{bmatrix} \tag{9.7}$$

$$\& \begin{bmatrix} 0 & 1 \\ -1 & 0 \end{bmatrix}^2 = \begin{bmatrix} -1 & 0 \\ 0 & -1 \end{bmatrix}$$

Also note that, when the real part, $a = 0$, the $\mathbb{C}$ matrix is an antisymmetric matrix. (The off-diagonal elements are the negatives of the transpose.)

---

[6] Sulla rappresentazione geometrica della algebra doppie dotate di modulo. Uldrico Bencivenga (1946).

[7] Calculus of Approximations. Bulletin de L'Academie Polonaise de Sciences (1956) Vol. 4 No.5 pp. 253–257.

[8] Complex numbersin geometry. Academic Press N.Y. pp. 18–20.

[9] Double and Dual numbers – Encyclopedia of Mathematics Soviet/AMS/Kluwer, Dordrect.

[10] Benz, W. Vorlesungen uber geometrie der algebren, Springer (1973).

[11] Extending Special Relativity with Perplex numbers. *American Journal of Physics* 54:416 (1986).

[12] Geometry of Lie groups, Kluwer academic publishers.

[13] Clifford Algebras and Spinors ISBN: 0-521-00551-5

[14] Complex numbers in n-dimensions ISBN: 0-444-51123-7

Throwing a minus sign into the C_2 matrix gives us the rotation matrix of 2-dimensional Euclidean space:

$$\exp\left(\begin{bmatrix} a & b \\ -b & a \end{bmatrix}\right) = \begin{bmatrix} e^a & 0 \\ 0 & e^a \end{bmatrix}\begin{bmatrix} 1 - \dfrac{b^2}{2!} + \dfrac{b^4}{4!} - \ldots & b - \dfrac{b^3}{3!} + \dfrac{b^5}{5!} - \ldots \\ b - \dfrac{b^3}{3!} + \dfrac{b^5}{5!} - \ldots & 1 - \dfrac{b^2}{2!} + \dfrac{b^4}{4!} - \ldots \end{bmatrix} \qquad (9.8)$$

$$= \begin{bmatrix} r & 0 \\ 0 & r \end{bmatrix}\begin{bmatrix} \cos b & \sin b \\ -\sin b & \cos b \end{bmatrix}$$

Actually, the "throwing in" of a minus sign is a little more than just "throwing in". The Euclidean complex numbers, $\mathbb{C}$, and the hyperbolic complex numbers, $\mathbb{S}$, (the space-time numbers) are essentially the only algebraic fields based on the finite group C_2, but the details are too far from the subject of this book to concern us. We now have the complex numbers, $\mathbb{C}$, and thus the 2-dimensional Euclidean space that is the complex plane (not $\mathbb{R}^2$) in addition to the hyperbolic complex numbers (2-dimensional space-time). There are no more geometric spaces inside C_2.

9.3   THE TWO TYPES OF 2-DIMENSIONAL SPACE

Because there are two and only two possible two-way splittings of the exponential series that differentiate into each other (the all pluses and the minus every second term splittings), there are two and only two types of 2-dimensional space – Euclidean and hyperbolic (space-time). This is why we have only two types of 2-dimensional rotations, spatial and space-time rotations. This is also why, in 2-dimensions, we have 2-way symmetry and anti-symmetry rather than some form of 3-way or 4-way symmetry.

So, from the finite group C_2, we have two 2-dimensional spaces. Our understanding of the 4-dimensional space-time which we seem to inhabit is that it contains three copies of each of these two 2-dimensional spaces. The six 2-dimensional rotations are perpendicular to each other. We call that construction of six 2-dimensional rotations the Lorentz group. Thus, the Lorentz group is made of two types of 2-dimensional rotations, rather than three or four types of 2-dimensional rotation, because of the nature of the exponential function. We will examine the Lorentz group in later chapters.

One might expect that an object moving through these two different spaces might be manifest in two different ways. One might expect that every object would have a kind of "duality" in how it is observed and that the observer would be able to influence how it is observed by changing the observing apparatus. Such "uncanny" behavior has been observed by physicists; they call it wave-particle duality.

Motion through the $\mathbb{C}$ plane is not motion through time – there is no time in the $\mathbb{C}$ plane. Thus, an object in the $\mathbb{C}$ plane can be at lots of different places at the same time. This is called non-locality. Non-locality is part of the uncanny nature of quantum mechanical phenomena.

Aside: It is by insisting on $U(1)$ local invariance that particle physicists derive the existence of the electromagnetic potential whose boson is the photon field (light). $U(1)$ invariance is invariance under rotation in the (Euclidean) complex plane. Can we conclude that light is a phenomenon of the Euclidean space that is the (Euclidean) complex plane? Of course, light proceeds at a speed that is outside of the hyperbolic complex number algebra (space-time).

9.4   TRIGONOMETRIC IDENTITIES

We have[15]:

$$\exp\left(\begin{bmatrix} a & b \\ -b & a \end{bmatrix}\right) = \exp\left(\begin{bmatrix} a & 0 \\ 0 & a \end{bmatrix}\right)\exp\left(\begin{bmatrix} 0 & b \\ b & 0 \end{bmatrix}\right) \qquad (9.9)$$

and that the determinate of the exponential of a matrix with zero trace is unity[16]. Thus

$$\det\left(\exp\left(\begin{bmatrix} 0 & b \\ -b & 0 \end{bmatrix}\right)\right) = \det\left(\begin{bmatrix} \cos b & \sin b \\ -\sin b & \cos b \end{bmatrix}\right) \qquad (9.10)$$

$$= \cos^2 b + \sin^2 b = 1$$

[15.] We can do this with matrices only if they commute, which these do.
[16.] This is a standard result (eigenvalues).

Which is our first trigonometric identity. Similarly:

$$\det\left(\exp\left(\begin{bmatrix} 0 & b \\ b & 0 \end{bmatrix}\right)\right) = \det\left(\begin{bmatrix} \cosh b & \sinh b \\ \sinh b & \cosh b \end{bmatrix}\right) \tag{9.11}$$

$$= \cosh^2 b - \sinh^2 b = 1$$

Notice the similarity

$$\cos^2 b + \sin^2 b = 1 \tag{9.12}$$

$$\cosh^2 b - \sinh^2 b = 1$$

between these two identities from two different types of space. The types of distances in these spaces are very different. One is what we normally call space, and the other is space-time. The angles in these spaces are very different. One is what we normally think of as an angle; the other is a change of velocity. Yet, there is a marked similarity between these spaces trigonometrically. Of course there is; they both derive from the group C_2, and, trigonometrically, they are both connected to two-way splittings of the exponential series.

Within 2-dimensional Euclidean space, we calculate the distance between two points by using the Pythagoras theorem. The Pythagoras theorem is not really anything to do with triangles; it is really the way that distance is calculated in 2-dimensional Euclidean space. There is a similar theorem, not necessarily quadratic, for each type of space. We have that the polar form of a complex number can be written in a Cartesian form. Thus:

$$\begin{bmatrix} r & 0 \\ 0 & r \end{bmatrix}\begin{bmatrix} \cos b & \sin b \\ -\sin b & \cos b \end{bmatrix} = \begin{bmatrix} x & y \\ -y & x \end{bmatrix} \tag{9.13}$$

Taking the determinant of both sides gives:

$$\det\left(\begin{bmatrix} r & 0 \\ 0 & r \end{bmatrix}\begin{bmatrix} \cos b & \sin b \\ -\sin b & \cos b \end{bmatrix}\right) = \det\left(\begin{bmatrix} x & y \\ -y & x \end{bmatrix}\right) \tag{9.14}$$

$$r^2 = x^2 + y^2$$

Which is the Pythagoras theorem (in Euclidean space). Further:

$$\det\left(\begin{bmatrix} h & 0 \\ 0 & h \end{bmatrix}\begin{bmatrix} \cosh b & \sinh b \\ \sinh b & \cosh b \end{bmatrix}\right) = \det\left(\begin{bmatrix} t & z \\ z & t \end{bmatrix}\right) \tag{9.15}$$

$$h^2 = t^2 - z^2$$

Which is the Pythagoras theorem (distance function) of space-time.

The product of two rotation matrices is the single matrix of a larger rotation:

$$\begin{bmatrix} \cos\theta & \sin\theta \\ -\sin\theta & \cos\theta \end{bmatrix}\begin{bmatrix} \cos\phi & \sin\phi \\ -\sin\phi & \cos\phi \end{bmatrix} \quad (9.16)$$

$$=\begin{bmatrix} \cos\theta\cos\phi - \sin\theta\sin\phi & \cos\theta\sin\phi + \sin\theta\cos\phi \\ -(\cos\theta\sin\phi + \sin\theta\cos\phi) & \cos\theta\cos\phi - \sin\theta\sin\phi \end{bmatrix}$$

$$=\begin{bmatrix} \cos(\theta+\phi) & \sin(\theta+\phi) \\ -\sin(\theta+\phi) & \cos(\theta+\phi) \end{bmatrix}$$

This is the origin of trigonometric identities like:

$$\cos(\theta+\phi) = \cos\theta\cos\phi - \sin\theta\sin\phi \quad (9.17)$$

And:

$$\begin{bmatrix} \cosh\theta & \sinh\theta \\ \sinh\theta & \cosh\theta \end{bmatrix}\begin{bmatrix} \cosh\phi & \sinh\phi \\ \sinh\phi & \cosh\phi \end{bmatrix}$$

$$=\begin{bmatrix} \cosh(\theta+\phi) & \sinh(\theta+\phi) \\ \sinh(\theta+\phi) & \cosh(\theta+\phi) \end{bmatrix} \quad (9.18)$$

Leading to:

$$\cosh(\theta+\phi) = \cosh\theta\cosh\phi + \sinh\theta\sinh\phi \quad (9.19)$$

and other similar hyperbolic trigonometric identities including:

$$\sinh(\theta+\phi) = \sinh\theta\cosh\phi + \cosh\theta\sinh\phi \quad (9.20)$$

The (Euclidean) tangent of an angle is the gradient of the line from the origin to the point on the circle. It is calculated as:

$$gradient = \tan\theta = \frac{\sin\theta}{\cos\theta} \quad (9.21)$$

There is a similar trigonometric function in space-time (the hyperbolic complex numbers). It is:

$$\tanh\chi = \frac{\sinh\chi}{\cosh\chi} \quad (9.22)$$

NOTE
$$\tanh(\theta+\phi)=\frac{\tanh\theta+\tanh\phi}{1+\tanh\theta\tanh\phi}$$

But, in space-time, the cosh( ) function is the projection on to the time axis and the sinh( ) function is the projection on to the space axis, and so:

$$\tanh\chi=\frac{\sinh\chi}{\cosh\chi}=\frac{\text{distance}}{\text{time}}=velocity \qquad (9.24)$$

That $\tanh\chi = velocity$ is an essential part of special relativity. Now:

$$\tanh\chi = v$$

$$\frac{\sinh^2\chi}{\cosh^2\chi}=v^2$$

$$1-\frac{1}{\cosh^2\chi}=v^2 \qquad (9.25)$$

$$\cosh\chi=\frac{1}{\sqrt{1-v^2}}$$

It follows that:

$$\sinh\chi=\frac{v}{\sqrt{1-v^2}} \qquad (9.26)$$

9.5   GAMMA

There is a minor complication that arises because we humans measure space and time in different units. We need to multiply space (the sinh( ) part) by the velocity of light, c, to balance the units of measurement that we humans use. When we do this, we get:

$$\cosh\chi=\frac{1}{\sqrt{1-\dfrac{v^2}{c^2}}}=\gamma=gamma$$

$$\qquad (9.27)$$

$$\sinh\chi=\frac{v}{\sqrt{1-\dfrac{v^2}{c^2}}}=v\gamma$$

The expression with the square root is universally[17] known as gamma, and is denoted by the Greek letter gamma. It occurs all over the place in special relativity. Now, cosh( ) is the projection from a unit circle (hyperbola) in space-time on to the time axis. Multiplying in space-time by cosh( ) effectively picks out the time part – this will become much clearer in the next chapter where it will be repeated. sinh( ) picks out the space part.

We have a plot of velocity against gamma (that passes through 1 on the vertical axis) and against the product of velocity and gamma (that passes through 0 on the vertical axis):

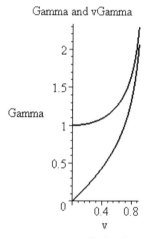

Gamma and vGamma

What we have done in space-time with the hyperbolic complex numbers, we can do in Euclidean space with the complex numbers. In this case, the ratio of the trigonometric functions $\tan\theta = \dfrac{\sin\theta}{\cos\theta}$ is the gradient of the line from the origin to the point on the circle and not the velocity:

$$\tan\theta = g$$

$$\frac{1-\cos^2\theta}{\cos^2\theta} = g^2 \qquad (9.28)$$

$$\cos\theta = \frac{1}{\sqrt{1+g^2}}$$

[17.] That is universally on Earth.

It follows that:

$$\sin\theta = \frac{g}{\sqrt{1+g^2}} \tag{9.29}$$

Because we measure horizontal space in the same way as we measure vertical space, there is no c^2 in this expression. We place the results side-by-side for the reader to cogitate upon:

$$\cos\theta = \frac{1}{\sqrt{1+g^2}} \quad : \quad \cosh\chi = \frac{1}{\sqrt{1-v^2}}$$

$$\sin\theta = \frac{g}{\sqrt{1+g^2}} \quad : \quad \sinh\chi = \frac{v}{\sqrt{1-v^2}} \tag{9.30}$$

We can now express our rotation matrices for Euclidean space and for space-time respectively as:

$$\frac{1}{\sqrt{1+g^2}}\begin{bmatrix} 1 & g \\ -g & 1 \end{bmatrix} \quad : \quad \frac{1}{\sqrt{1-v^2}}\begin{bmatrix} 1 & v \\ v & 1 \end{bmatrix} = \begin{bmatrix} \gamma & v\gamma \\ v\gamma & \gamma \end{bmatrix} \tag{9.31}$$

Since time and space are the same things, we ought to balance the nature of the elements of the space-time rotation matrix as:

$$\begin{bmatrix} \gamma & \dfrac{v}{c}\gamma \\ \dfrac{v}{c}\gamma & \gamma \end{bmatrix} \tag{9.32}$$

The determinant of these rotation matrices is unity; of course it is – they are rotation matrices:

$$\det\left(\begin{bmatrix} \gamma & v\gamma \\ v\gamma & \gamma \end{bmatrix}\right) = \gamma^2 - v^2\gamma^2 = \gamma^2(1-v^2) \tag{9.33}$$

$$= \left(\frac{1}{\sqrt{1-v^2}}\right)^2(1-v^2) = 1$$

This means that, since, for matrices in general $\det(A)\det(B) = \det(AB)$, multiplying a matrix by the rotation matrix will leave the determinant unchanged. The determinant is an invariant of the rotation. The determinant is the distance function. It is the distance from the origin that is invariant. That is what rotation is!

Aside: Within general relativity, within a gravitational field, we have:

$$\gamma = \frac{1}{\sqrt{1 - \frac{v^2}{c^2} + \frac{a}{r}}} \tag{9.34}$$

The $\frac{a}{r}$ term is the gravitational potential term.

9.6   THE GRAPHS OF THE TRIGONOMETRIC FUNCTIONS

The nature of the Euclidean trigonometric functions is that the $\sin(\ )$ function is just the $\cos(\ )$ function displaced by $\frac{\pi}{2}$. The graphs of these functions are:

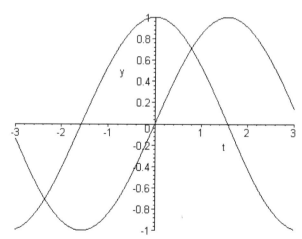

It is because these functions repeat themselves every 2π that we are able to rotate all the way round to where we started after 2π. Because $\sin\left(\theta - \frac{\pi}{2}\right) = \cos(\theta)$, the two axes of the complex plane are of the same spatial nature, but rotated by $\frac{\pi}{2}$. There is no such relation between the hyperbolic trigonometric functions. The hyperbolic trigonometric functions do not repeat themselves. The graphs of {cosh( ), sinh( )} are:

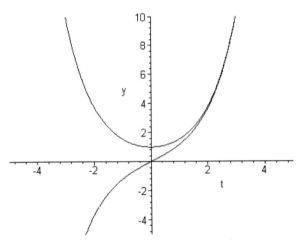

wherein the graphs fly off to ± infinity in the vertical direction. The cosh( ) graph is the one that does not cross the horizontal axis. Note that, in contradistinction to the cosine function of Euclidean space, $\cosh \chi = \gamma \geq 1$.

Although it is difficult to see, we have that sinh( ) is less than cosh( ) for all values. This is important because it means that the hyperbolic complex numbers are a (handed) division algebra. The matrix:

$$\begin{bmatrix} h & 0 \\ 0 & h \end{bmatrix} \begin{bmatrix} \cosh b & \sinh b \\ \sinh b & \cosh b \end{bmatrix} \tag{9.35}$$

will never have a zero determinant, and so every such matrix has a multiplicative inverse. That sinh( ) < cosh( ) also means that the velocity, $v = \dfrac{\sinh \chi}{\cosh \chi}$ will always be less than 1 (that is always less than the speed of light). It is the limiting velocity of space-time that ensures the order of events is conserved (cause and effect) – see later. This limiting velocity exists because of the nature of the hyperbolic trigonometric functions. They inherit this nature from the exponential function. Thus, we have the exponential function to thank for our living in a "cause and effect" universe.

If the universe allowed the sinh( ) function to equal the cosh( ) function, then the hyperbolic complex numbers would not be a division algebra (singular matrices would be possible). If the universe allowed the sinh( ) function to exceed the cosh( ) function, then there would be no ordering of events in the universe and thus no cause and effect in the universe. The universe does exactly as much as is possible without upsetting folk.

Note that the cosh( ) function is never zero (look at the series expansion above). Thus the projection from space-time on to the time axis can never be zero. This is why time never stops flowing. It seems that it might also be why we have non-zero rest mass and non-zero electric charge (electrons).

To repeat: Inspection of the series expansions of the cosh( ) and sinh( ) functions will show a differentiation cycle:

$$\frac{d}{dx}\cosh x = \sinh x, \quad \frac{d}{dx}\sinh x = \cosh x \qquad (9.36)$$

Similarly for {cos( ), sin( )}

$$\frac{d}{dx}\cos x = -\sin x, \quad \frac{d}{dx}\sin x = \cos x \qquad (9.37)$$

Since these functions are projections on to the different axes, the differentiation cycle "defines" a unit circle to be such that the horizontal co-ordinate is the rate of change with respect to the angle of the vertical co-ordinate (up to a sign) and the vertical co-ordinate is the rate of change of the horizontal co-ordinate with respect to the angle. This coincides with the geometric definition of a circle as being a set of points at unit distance from the origin.

Let us differentiate the rotation matrix with respect to the angle. We use the off-diagonal matrix element because it is this that we exponentiate to get the rotation matrix. The reader might be unfamiliar with matrix differentiation; we cover it later:

$$\frac{d\begin{bmatrix} \cosh\chi & \sinh\chi \\ \sinh\chi & \cosh\chi \end{bmatrix}}{d\begin{bmatrix} 0 & \chi \\ \chi & 0 \end{bmatrix}} = \frac{d\begin{bmatrix} \cosh\chi & \sinh\chi \\ \sinh\chi & \cosh\chi \end{bmatrix}}{\begin{bmatrix} 0 & 1 \\ 1 & 0 \end{bmatrix}d\begin{bmatrix} \chi & 0 \\ 0 & \chi \end{bmatrix}}$$

$$= \frac{1}{\begin{bmatrix} 0 & 1 \\ 1 & 0 \end{bmatrix}}\begin{bmatrix} \sinh\chi & \cosh\chi \\ \cosh\chi & \sinh\chi \end{bmatrix} \tag{9.38}$$

$$= \begin{bmatrix} 0 & 1 \\ 1 & 0 \end{bmatrix}\begin{bmatrix} \sinh\chi & \cosh\chi \\ \cosh\chi & \sinh\chi \end{bmatrix}$$

$$= \begin{bmatrix} \cosh\chi & \sinh\chi \\ \sinh\chi & \cosh\chi \end{bmatrix}$$

And so, the rotation matrix differentiated with respect to the angle is the rotation matrix. Note that the rotation matrix is derived from the $\begin{bmatrix} 0 & z \\ z & 0 \end{bmatrix}$ part of the hyperbolic complex number, and so we have to take account of this when we differentiate.

Note to sci-fi writers: To construct a universe, first construct, or discover, a space. Calculate the rotation matrix in that space and then find the physical laws that would be invariant under rotation in that space. You now have the basis for a well founded sci-fi story!

PS: The finite group C_3 or, for the ambitious, A_5 will give you some true sci-fi.

EXERCISES

1. Force from empty spaces: Use the Euclidean distance function (Pythagoras) to draw a straight line passing through three points in the Euclidean plane (that is on a flat piece of paper). Now, and this is not easy, draw the same line passing through the same points in space-time (with the space-time distance function). Is the line is space-time straight? (It is

not straight.) Now, an observer in Euclidean space watching an object move in a straight line through Euclidean space will say that no force acts upon that object. Will an observer in space-time, who sees the same object move in a curved trajectory, see a force acting upon that object?

2. Calculate to three terms $\sin\left(\dfrac{\pi}{4}\right)$ from the series (9.3).

3. Calculate to three terms $\cosh(ix)$ from the series (9.2) and compare it to the series for the $\cos(\ )$ function.

4. Calculate to three terms $\sinh(ix)$ from the series (9.2) and compare it to the series for the $\sin(\ )$ function.

5. Differentiate the infinite series $\dfrac{x^2}{2!}+\dfrac{x^5}{5!}+\dfrac{x^8}{8!}+...$ with respect to x and compare it to (9.4).

6. (Just for fun) Sum the series in (9.4) and calculate the product of the two matrices. The answer will surprise you!

10

INTRODUCTION TO VECTORS

The temperature at each position in a room is a real number associated with that position. We say that the temperature at the point whose co-ordinates are, say, (1, 2, 3) is the number, say, 35. This means that, at that point, the temperature is 35°C. At a different point, the number might be 32. Such a distribution of real numbers, one for each point in the room, is called a scalar field. A scalar field is just a single real number at each point.

Now, some things, and a magnetic field is one of these things, have, not only a numerical value at each point in a room, but also a direction – think direction pointed by a compass needle. The magnetic field might have strength 2 in the down direction at one point in the room and strength 5 in the up direction at a different point in the room. Such a field is called a vector field. A vector field is a strength and a direction at each co-ordinate point. We visualise a vector field as being a set of little arrows with one little arrow being at each point in the space. The direction of each arrow indicates the direction of the vector field at that point, and the length of each arrow is equal to the field strength at that point.

Aside: Vectors were first introduced by Hermann Grassman (1809–1877), but vector calculus was invented by Josiah Willard Gibbs (1839–1903) and Oliver Heaviside (1850–1925) in 1888 when they rewrote Maxwell's quaternion formulation of electromagnetism in vector form. The dot product and the cross product were invented by Gibbs in 1903.

A vector in a space (or space-time) is a set of ordered real numbers. There is one number for each dimension of the space (or space-time). The ratios of the numbers give the direction of the vector and the values of the numbers allow us to calculate the length (strength) of the vector. For example: conventionally, a vector in 2-dimensional space looks like:

$$\begin{bmatrix} 2 \\ 1 \end{bmatrix} \tag{10.1}$$

The top number (the 2) is the length of the vector in the horizontal direction (the x co-ordinate), and the bottom number is the length of the vector in the vertical direction (the y co-ordinate). The whole vector is the arrow starting at the origin and ending at the point (2, 1):

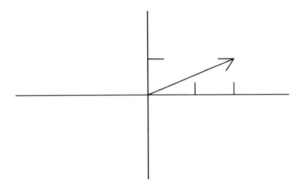

Vectors are often written in the form: $a\vec{e_1} + b\vec{e_2}$ where $\{\vec{e_1}, \vec{e_2}\}$ are the orthogonal unit basis vectors along each axis of length one in the different directions. The above vector would be written in this form as: $2\vec{e_1} + \vec{e_2}$.

We can work out the length of the vector using the Pythagoras theorem. In Euclidean space, this is:

$$length = \sqrt{x^2 + y^2} = \sqrt{2^2 + 1^2} = \sqrt{5} \tag{10.2}$$

In space-time, the Pythagorean like theorem is:

$$length = \sqrt{t^2 - z^2} = \sqrt{2^2 - 1^2} = \sqrt{3} \tag{10.3}$$

It is important to realize that each type of space has its own type of distance function (Pythagoras type theorem). However, in all types of space, a vector is just an ordered set of real numbers that are the co-ordinates of the head of the arrow representing the vector.

Above, we have chosen a co-ordinate system that has the x-axis horizontal and the y-axis vertical. We need not have chosen them this way. We might have chosen that the x-axis was at 45° to the horizontal and the y-axis similarly adjusted. However, the vector represents a thing, like a magnetic field, that really exists. Magnetic fields do not adjust themselves to suit how we arbitrarily choose to align our co-ordinate axes. The vector does not change its direction because we have changed our co-ordinate system. If we did choose our co-ordinate system such that the x-axis pointed at 45° to the horizontal, then, in 2-dimensional Euclidean space, the vector would be the "rotated" numbers:

$$\begin{bmatrix} \cos\dfrac{\pi}{4} & \sin\dfrac{\pi}{4} \\ -\sin\dfrac{\pi}{4} & \cos\dfrac{\pi}{4} \end{bmatrix}\begin{bmatrix} 2 \\ 1 \end{bmatrix} = \begin{bmatrix} 2.\cos\dfrac{\pi}{4}+1.\sin\dfrac{\pi}{4} \\ -2.\sin\dfrac{\pi}{4}+1.\cos\dfrac{\pi}{4} \end{bmatrix} = \begin{bmatrix} \dfrac{3}{\sqrt{2}} \\ -\dfrac{1}{\sqrt{2}} \end{bmatrix} \quad (10.4)$$

Note that the 2-dimensional Euclidean rotation matrix changes the components of the vector appropriately. The components of the vector change when we change the co-ordinate system, which ought not to surprise the reader, but the (Euclidean) length of the vector stays the same:

$$\begin{aligned} length &= \sqrt{\left(2.\cos\dfrac{\pi}{4}+1.\sin\dfrac{\pi}{4}\right)^2 + \left(-2.\sin\dfrac{\pi}{4}+1.\cos\dfrac{\pi}{4}\right)^2} \\ &= \sqrt{4\left(\cos^2\dfrac{\pi}{4}+\sin^2\dfrac{\pi}{4}\right)+1\left(\cos^2\dfrac{\pi}{4}+\sin^2\dfrac{\pi}{4}\right)} \\ &= \sqrt{5} \end{aligned} \quad (10.5)$$

Which ought not to surprise the reader until they think about how remarkable it really is that a rotation matrix from the finite group C_2 preserves the length of a vector under rotation of co-ordinate system. It works because $\cos^2\theta + \sin^2\theta = 1$, and this is the case because the determinant of the exponential of a matrix with zero trace is always unity; it is by exponentiating such a matrix that we derived the rotation matrix, and it is from C_2 that we have such a matrix. Does it not fit together well?

In space-time:

$$\begin{bmatrix} \cosh\dfrac{\pi}{4} & \sinh\dfrac{\pi}{4} \\ \sinh\dfrac{\pi}{4} & \cosh\dfrac{\pi}{4} \end{bmatrix} \begin{bmatrix} 2 \\ 1 \end{bmatrix} = \begin{bmatrix} 2.\cosh\dfrac{\pi}{4}+1.\sinh\dfrac{\pi}{4} \\ 2.\sinh\dfrac{\pi}{4}+1.\cosh\dfrac{\pi}{4} \end{bmatrix} \qquad (10.6)$$

With length:

$$\begin{aligned} length &= \sqrt{\left(2.\cosh\dfrac{\pi}{4}+1.\sinh\dfrac{\pi}{4}\right)^2 - \left(2.\sinh\dfrac{\pi}{4}+1.\cosh\dfrac{\pi}{4}\right)^2} \\ &= \sqrt{4\left(\cosh^2\dfrac{\pi}{4}-\sinh^2\dfrac{\pi}{4}\right)-1\left(\cosh^2\dfrac{\pi}{4}-\sinh^2\dfrac{\pi}{4}\right)} \\ &= \sqrt{3} \end{aligned} \qquad (10.7)$$

We have written earlier in this book about the physics of the universe being invariant under change of direction. We pointed out that kettles boil at the same temperature regardless of the direction in which their spouts point or the speed at which they move. We have just seen that the length of a vector, which is the strength of the vector (magnetic) field, does not vary when we rotate our co-ordinate system. This is the same thing as the invariance under rotation of physical phenomena, but we are rotating the co-ordinate system rather than the physical apparatus.

Now, how we orientate our co-ordinate system, and whether we choose a Cartesian co-ordinate system or some other co-ordinate system like a polar co-ordinate system for example, is an arbitrary choice made by humankind. It seems reasonable that the physical phenomena of the universe should be the same regardless of how we choose our co-ordinate system. So, what is there about a vector field that is invariant under change of co-ordinate system? The components of a vector vary with the co-ordinate system chosen, but the length of the vector stays the same regardless of how we change the co-ordinate system. Since vector length corresponds to the field strength, this means that the field strength stays the same under change of co-ordinate system.

As well as the length of a vector, there is one other thing about a vector field that is invariant under change of co-ordinates. This is the angle between two vectors; this effectively means that the direction of the vector field stays the same under change of co-ordinate system.

We have written a vector above in the conventional way as a column matrix of two components. We then multiplied this column matrix by a (square) rotation matrix[1]. The reader will have been taught that it is okay to multiply a column matrix by a square matrix, and mathematicians and physicists do this all the time, but, to a pedant, this is wrong. To be perfectly correct, one cannot multiply a dog by a cat, and one cannot multiply a square matrix by a not-square matrix. This is not a problem to us because we are now going to write our vectors as square matrices. A vector in 2-dimensional Euclidean space (the complex plane, $\mathbb{C}$) is just a complex number and is written as:

$$\begin{bmatrix} u \\ v \end{bmatrix} \equiv \begin{bmatrix} u & v \\ -v & u \end{bmatrix} = \begin{bmatrix} r & 0 \\ 0 & r \end{bmatrix} \begin{bmatrix} \cos\theta & \sin\theta \\ -\sin\theta & \cos\theta \end{bmatrix} \tag{10.8}$$

We note that the length of the vector:

$$\begin{bmatrix} 2 & 1 \\ -1 & 2 \end{bmatrix} \tag{10.9}$$

is just the square root of the determinant: $\sqrt{2^2 + 1^2} = \sqrt{5}$. Of course, the determinant of a matrix is invariant under change of basis of the matrix (similarity transformation).

A vector in 2-dimensional space-time is just a hyperbolic complex number and is written as:

$$\begin{bmatrix} t \\ z \end{bmatrix} \equiv \begin{bmatrix} t & z \\ z & t \end{bmatrix} = \begin{bmatrix} h & 0 \\ 0 & h \end{bmatrix} \begin{bmatrix} \cosh\chi & \sinh\chi \\ \sinh\chi & \cosh\chi \end{bmatrix} \tag{10.10}$$

We note that the length of the vector:

$$\begin{bmatrix} 2 & 1 \\ 1 & 2 \end{bmatrix} \tag{10.11}$$

is just the square root of the determinant:

$$\sqrt{2^2 - 1^2} = \sqrt{3} \tag{10.12}$$

---

[1] All rotation matrices are square – it is not hard to square a circle.

It is now much easier for us to calculate the angle between two vectors. Let us take two vectors in the $\mathbb{C}$ plane. Let the angle between them be ϕ. In polar form, these vectors are:

$$\begin{bmatrix} r & 0 \\ 0 & r \end{bmatrix} \begin{bmatrix} \cos\theta & \sin\theta \\ -\sin\theta & \cos\theta \end{bmatrix} \quad \& $$
$$\begin{bmatrix} s & 0 \\ 0 & s \end{bmatrix} \begin{bmatrix} \cos(\theta+\phi) & \sin(\theta+\phi) \\ -\sin(\theta+\phi) & \cos(\theta+\phi) \end{bmatrix} \tag{10.13}$$

First we normalize these vectors by dividing by $\{r, s\}$ respectively. Then we take the conjugate of one (either will do – watch the minus sign) and multiply the two together.

$$\begin{bmatrix} \cos\theta & -\sin\theta \\ \sin\theta & \cos\theta \end{bmatrix} \begin{bmatrix} \cos(\theta+\phi) & \sin(\theta+\phi) \\ -\sin(\theta+\phi) & \cos(\theta+\phi) \end{bmatrix} \tag{10.14}$$
$$= \begin{bmatrix} \cos(-\phi) & -\sin(-\phi) \\ \sin(-\phi) & \cos(-\phi) \end{bmatrix}$$
$$= \begin{bmatrix} \cos\phi & \sin\phi \\ -\sin\phi & \cos\phi \end{bmatrix}$$

We have an expression for the angle, ϕ, between the two vectors. We will do exactly the same with the Cartesian forms of the vectors. Let the two vectors in Cartesian form be:

$$\begin{bmatrix} a & b \\ -b & a \end{bmatrix} \quad \& \quad \begin{bmatrix} c & d \\ -d & c \end{bmatrix} \tag{10.15}$$

Normalizing, taking the conjugate of the leftmost, and multiplying gives:

$$\begin{bmatrix} \dfrac{a}{\sqrt{a^2+b^2}} & -\dfrac{b}{\sqrt{a^2+b^2}} \\ \dfrac{b}{\sqrt{a^2+b^2}} & \dfrac{a}{\sqrt{a^2+b^2}} \end{bmatrix} \begin{bmatrix} \dfrac{c}{\sqrt{c^2+d^2}} & \dfrac{d}{\sqrt{c^2+d^2}} \\ -\dfrac{d}{\sqrt{c^2+d^2}} & \dfrac{c}{\sqrt{c^2+d^2}} \end{bmatrix}$$

$$\tag{10.16}$$

$$= \begin{bmatrix} \dfrac{ac+bd}{\sqrt{a^2+b^2}\sqrt{c^2+d^2}} & \dfrac{ad-bc}{\sqrt{a^2+b^2}\sqrt{c^2+d^2}} \\ -\left(\dfrac{ad-bc}{\sqrt{a^2+b^2}\sqrt{c^2+d^2}}\right) & \dfrac{ac+bd}{\sqrt{a^2+b^2}\sqrt{c^2+d^2}} \end{bmatrix}$$

Putting these two together gives:

$$\begin{bmatrix} \cos\phi & \sin\phi \\ -\sin\phi & \cos\phi \end{bmatrix} = \begin{bmatrix} \dfrac{ac+bd}{\sqrt{a^2+b^2}\sqrt{c^2+d^2}} & \dfrac{ad-bc}{\sqrt{a^2+b^2}\sqrt{c^2+d^2}} \\ -\left(\dfrac{ad-bc}{\sqrt{a^2+b^2}\sqrt{c^2+d^2}}\right) & \dfrac{ac+bd}{\sqrt{a^2+b^2}\sqrt{c^2+d^2}} \end{bmatrix} \quad (10.17)$$

And so, we can calculate the angle between the two vectors as either the cos( ) of the angle or the sin( ) of the angle from the Cartesian components. The expression:

$$ac+bd \equiv \begin{bmatrix} a \\ b \end{bmatrix} \bullet \begin{bmatrix} c \\ d \end{bmatrix} \equiv \left\langle \begin{bmatrix} a \\ b \end{bmatrix}, \begin{bmatrix} c \\ d \end{bmatrix} \right\rangle \quad (10.18)$$

is called the dot product of the two vectors in Euclidean space. It is also called the inner product of the vectors. Note that:

$$\begin{bmatrix} a \\ b \end{bmatrix} \bullet \begin{bmatrix} c \\ d \end{bmatrix} = \begin{bmatrix} c \\ d \end{bmatrix} \bullet \begin{bmatrix} a \\ b \end{bmatrix} \quad (10.19)$$

This is because the angle ϕ between two vectors is the same when measured "clockwise" as it is when measured "counterclockwise". This "obvious" fact would not be the case if it were not that the cos( ) function is symmetrical about the vertical axis (see graph). For this, we thank the exponential function; without it, an angle measured in the clockwise direction would be different from the same angle measured in the counterclockwise direction.

The expression:

$$ad-bc \equiv \begin{bmatrix} a \\ b \end{bmatrix} \times \begin{bmatrix} c \\ d \end{bmatrix} \quad (10.20)$$

is called the cross-product, or outer product, of the two vectors in Euclidean space. The cross-product of two vectors is often thought of as an axial vector in 3-dimensional space. The idea is that the cross-product of two vectors in 3-dimensional space is a vector that is perpendicular to the plane that contains the two vectors that are crossed together to form the cross-product. There is no place for such a notion in 2-dimensional space. We are working in the 2-dimensional spaces that are derived from the finite group C_2. These spaces cannot "grow" another dimension to accommodate the notion of the cross-product any more than the finite group C_2 can transmute into the finite group C_3. In these spaces, the cross-product is simply another way to calculate the angle between two vectors. However, the cross-product does occur often in classical physics (the electromagnetic force law), and the reader will need to know of it, and of its interpretation as a vector perpendicular to the plane of the two component vectors that form it in $\mathbb{R}^3$ space.

Aside: Since Grassman's first work with vectors, mathematicians have sought to introduce a way of multiplying two vectors together to give a vector. Grassman introduced the exterior product and William Kingdon Clifford (1845–1879) introduced the Clifford product[2], neither of which really work. The Grassman algebras have non-zero numbers that square to zero, and the Clifford product is a scalar and a vector. In two dimensions, cl_2, the Clifford product is based on:

$$\left(\vec{e}_1\right)^2 = \left(\vec{e}_2\right)^2 = 1 \quad \& \quad \vec{e}_1\vec{e}_2 = -\vec{e}_2\vec{e}_1 \qquad (10.21)$$

$$\left(\vec{e}_1\vec{e}_2\right)^2 = \vec{e}_1\vec{e}_2\vec{e}_1\vec{e}_2 = -\left(\vec{e}_1\right)^2\left(\vec{e}_2\right)^2 = -1$$

Thus:

$$\left(a_1\vec{e}_1 + a_2\vec{e}_2\right)\left(b_1\vec{e}_1 + b_2\vec{e}_2\right) = a_1b_1\left(\vec{e}_1\right)^2 + a_1b_2\vec{e}_1\vec{e}_2$$

$$+ a_2b_1\vec{e}_2\vec{e}_1 + a_2b_2\left(\vec{e}_2\right)^2 \qquad (10.22)$$

$$= a_1b_1 + a_2b_2 + \left(a_1b_2 - a_2b_1\right)\vec{e}_1\vec{e}_2$$

Wherein we effectively have both the dot product and the cross product. We note that we have two square roots of plus one and one square root of minus one. The reader might think we have a type of complex number. With a minor change, we do have a type of complex number; this is an A_3 algebra.[3]

Aside: The cross-product of two vectors is dual to a vector in only $\mathbb{R}^3$ space and $\mathbb{R}^7$ space. This notion has its origin in the Clifford algebras. This is connected to the existence of the quaternions and the octonians. The cross-product in $\mathbb{R}^4$ is seen as a 2-dimensional plane, and the cross product in $\mathbb{R}^5$ is seen as a 3-dimensional volume.

[2.] Applications of Grassman's extensive algebra American J., 1, 350–358 M. P. 266–276.
[3.] See: Dennis Morris : The Naked Spinor – a rewrite of Clifford algebra ISBN: 978-1507817995

If we use the dot product to calculate the angle between a vector and itself (which is zero), we get:

$$\begin{bmatrix} \cos 0 & \sin 0 \\ -\sin 0 & \cos 0 \end{bmatrix} = \begin{bmatrix} \dfrac{a^2 + b^2}{\sqrt{a^2 + b^2}\sqrt{a^2 + b^2}} & \dfrac{ab - ba}{\sqrt{a^2 + b^2}\sqrt{a^2 + b^2}} \\ -\left(\dfrac{ab - ba}{\sqrt{a^2 + b^2}\sqrt{a^2 + b^2}}\right) & \dfrac{a^2 + b^2}{\sqrt{a^2 + b^2}\sqrt{a^2 + b^2}} \end{bmatrix} \quad (10.23)$$

$$= \begin{bmatrix} 1 & 0 \\ 0 & 1 \end{bmatrix}$$

But if we do not normalize the vector, we get:

$$r^2 = a^2 + b^2 \equiv \begin{bmatrix} a \\ b \end{bmatrix} \bullet \begin{bmatrix} a \\ b \end{bmatrix} \equiv \left\langle \begin{bmatrix} a \\ b \end{bmatrix}, \begin{bmatrix} a \\ b \end{bmatrix} \right\rangle \quad (10.24)$$

Thus, the dot product of a vector with itself is the length (squared in 2-dimensional space) of the vector.

To calculate the angle between two vectors in space-time, we do the same as above but adjusted for the different type of space.

$$\begin{bmatrix} \cosh(\varphi) & \sinh(\varphi) \\ \sinh(\varphi) & \cosh(\varphi) \end{bmatrix} \bullet \begin{bmatrix} \cosh(\varphi - \chi) & \sinh(\varphi - \chi) \\ \sinh(\varphi - \chi) & \cosh(\varphi - \chi) \end{bmatrix}$$

$$= \begin{bmatrix} \cosh(\varphi) & \sinh(\varphi) \\ \sinh(\varphi) & \cosh(\varphi) \end{bmatrix} \begin{bmatrix} \cosh(\varphi - \chi) & -\sinh(\varphi - \chi) \\ -\sinh(\varphi - \chi) & \cosh(\varphi - \chi) \end{bmatrix} \quad (10.25)$$

$$\begin{bmatrix} a & b \\ b & a \end{bmatrix} \bullet \begin{bmatrix} c & d \\ d & c \end{bmatrix} = \begin{bmatrix} a & b \\ b & a \end{bmatrix} \begin{bmatrix} c & -d \\ -d & c \end{bmatrix}$$

This leads to:

$$\begin{bmatrix} \cosh \chi & \sinh \chi \\ \sinh \chi & \cosh \chi \end{bmatrix} = \begin{bmatrix} \dfrac{ac - bd}{\sqrt{a^2 - b^2}\sqrt{c^2 - d^2}} & \dfrac{ad - bc}{\sqrt{a^2 - b^2}\sqrt{c^2 - d^2}} \\ \dfrac{ad - bc}{\sqrt{a^2 - b^2}\sqrt{c^2 - d^2}} & \dfrac{ac - bd}{\sqrt{a^2 - b^2}\sqrt{c^2 - d^2}} \end{bmatrix} \quad (10.26)$$

Of course, the space-time angle, χ, between two vectors is a difference in velocities. The length of a vector is given by the dot product of the vector with itself:

$$h^2 = a^2 - b^2 \equiv \begin{bmatrix} a \\ b \end{bmatrix} \bullet \begin{bmatrix} a \\ b \end{bmatrix} \equiv \left\langle \begin{bmatrix} a \\ b \end{bmatrix}, \begin{bmatrix} a \\ b \end{bmatrix} \right\rangle \quad (10.27)$$

In the case of space-time, the algebra of the hyperbolic complex numbers demands that the real part of a vector be greater than the imaginary part (except for the zero vector) to avoid singular matrices, and so the length of the vector must be positive.

We can see that the dot product of two vectors is different for each type of space, but the notation is, conventionally, the same. We will have to be careful.

10.1 ORTHOGONALITY AND PERPENDICULARITY

In the Euclidean space, the cosine function is zero when the angle is 90°. Since the normalized dot product of two vectors is equal to the cosine of the angle between them, when the dot product of two vectors is zero, the angle between the vectors is 90° – the two vectors are perpendicular to each other. In Euclidean space, when the dot product of two vectors is zero, we say that the two vectors are orthogonal. In Euclidean space, orthogonal means the same as perpendicular.

It is also conventional to say that two vectors in space-time are orthogonal if their dot product is zero; however, this is not without a difficulty. We have:

$$\begin{bmatrix} a & b \\ b & a \end{bmatrix} \bullet \begin{bmatrix} c & d \\ d & c \end{bmatrix} = \begin{bmatrix} a & b \\ b & a \end{bmatrix} \begin{bmatrix} c & -d \\ -d & c \end{bmatrix}$$

$$\begin{bmatrix} ac - bd & bc - ad \\ bc - ad & ac - bd \end{bmatrix} = \begin{bmatrix} \cosh \chi & \sinh \chi \\ \sinh \chi & \cosh \varphi \end{bmatrix}$$

(10.28)

The cosh( ) function can never be zero, and so the dot product is never zero (unless the vector is of zero length, which doesn't count) – that's the difficulty. Like the cos( ) function is unity if the angle is zero, so too the cosh( ) function is unity if the angle is zero, but there is no angle in space-time that makes the cosh( ) function zero. This fits with the algebraic requirement that, except for the zero matrix, the absolute value of the real part of a hyperbolic complex number must be greater than the absolute value of the imaginary part – in the above we have: $\{a > b, c > d\} \Rightarrow ac - bd > 0$. Thus, the convention of saying two vectors are orthogonal in space-time if their dot product is zero is nonsense – that's the difficulty, again. This does not mean that the concept of perpendicularity is meaningless in space-time.

In Euclidean space, perpendicular means being at 90°. We can take the view that "perpendicular in a general sense" means "independent of". In which case, perpendicular means that the x-direction is independent of the y-direction, and the same applies to space-time with the time-axis being independent of the space-axis.

We will later find that we must adopt the view that orthogonal means "in an independent direction in some basis". Only with this view is it sensible to say that, for Euclidean space, the x-axis is orthogonal to the y-axis, and, for space-time, the time-axis is orthogonal to the space-axis. It is "accidental" that within Euclidean space the dot product is zero exactly when the argument of the cos( ) function is 90°. Such an "accident" cannot happen in space-time.

10.2 DIFFERENTIATION (THE STANDARD VIEW)

In Newtonian mechanics, when we want to know the velocity of an object, we differentiate its distance with respect to time:

$$v = \frac{ds}{dt} \tag{10.29}$$

If we want to know the object's acceleration, we differentiate its velocity with respect to time:

$$a = \frac{dv}{dt} = \frac{d^2s}{dt^2} \tag{10.30}$$

This works in Newtonian mechanics because we assume that the processes of the universe flow at the same rate for both the stationary observer and the moving object being observed. Except for very high velocities, this is very close to being true. However, since, as we will see in more detail later, the processes of the universe flow at a slower rate for a moving object than for a stationary observer, when we differentiate with respect to time, we are differentiating with respect to something that is itself varying. We can do this provided we know how the variable with respect to which we are differentiating is varying, but it is conceptually easier to differentiate with respect to something that is invariant. For a single vector, the only invariant is its length, which is the determinant of the matrix that represents it. Thus, we need to differentiate with respect to the determinant. We will do this for space-time; we will denote the determinant by the squared Greek letter τ. We have:

$$\tau^2 = t^2 - z^2 = \det\left(\begin{bmatrix} t & z \\ z & t \end{bmatrix}\right) \quad : \quad |t| > |z| \tag{10.31}$$

We take the differentials:

$$(d\tau)^2 = (dt)^2 - (dz)^2$$

$$\Rightarrow \frac{d\tau}{dt} = \sqrt{1 - \left(\frac{dz}{dt}\right)^2} = \sqrt{1 - v^2} \tag{10.32}$$

$$\Rightarrow \frac{dt}{d\tau} = \frac{1}{\sqrt{1 - v^2}}$$

Inserting the velocity of light, c, to adjust the units of space to time, gives:

$$\frac{dt}{d\tau} = \frac{1}{\sqrt{1 - \dfrac{v^2}{c^2}}} = \gamma = \cosh \chi \tag{10.33}$$

It's that gamma thing again!

Again, we take the differentials:

$$(d\tau)^2 = (dt)^2 - (dz)^2$$

$$\Rightarrow \frac{d\tau}{dz} = \sqrt{\left(\frac{dt}{dz}\right)^2 - 1} = \sqrt{\frac{1}{v^2} - 1} = \frac{1}{v}\sqrt{1 - v^2} \tag{10.34}$$

$$\Rightarrow \frac{dz}{d\tau} = \frac{v}{\sqrt{1 - v^2}}$$

Inserting the velocity of light, c, to adjust the units of space to time, gives:

$$\frac{dz}{d\tau} = \frac{v}{\sqrt{1 - \dfrac{v^2}{c^2}}} = v\gamma = \sinh \chi \tag{10.35}$$

This is that sinh( ) thing again.

We do the same as above with Euclidean space. This time we will denote the determinant by the squared Greek letter σ.

$$\sigma^2 = x^2 + y^2 = \det\left(\begin{bmatrix} x & y \\ -y & x \end{bmatrix}\right) \tag{10.36}$$

We take the differentials:

$$(d\sigma)^2 = (dx)^2 + (dy)^2$$

$$\Rightarrow \frac{d\sigma}{dx} = \sqrt{1 + \left(\frac{dy}{dx}\right)^2} = \sqrt{1 + g^2} \tag{10.37}$$

$$\Rightarrow \frac{dx}{d\sigma} = \frac{1}{\sqrt{1 + g^2}} = \cos\theta$$

We have no need to adjust the units. Similarly, we have:

$$\frac{dy}{d\sigma} = \frac{g}{\sqrt{1 + g^2}} = \sin\theta \tag{10.38}$$

10.3 DIFFERENTIATION WITH MATRICES

We begin with the space-time matrix in 2-dimensional space-time. This is:

$$\begin{bmatrix} t & z \\ z & t \end{bmatrix} \quad : \quad |t| > |z| \tag{10.39}$$

We differentiate it with respect to the invariant interval, τ. Now τ is a real number, and so it is of the form:

$$\begin{bmatrix} \tau & 0 \\ 0 & \tau \end{bmatrix} \tag{10.40}$$

Differentiating with respect to this matrix is just differentiation with respect to a real number. We have:

$$\frac{d\begin{bmatrix} t & z \\ z & t \end{bmatrix}}{d\begin{bmatrix} \tau & 0 \\ 0 & \tau \end{bmatrix}} = \begin{bmatrix} \dfrac{dt}{d\tau} & \dfrac{dz}{d\tau} \\ \dfrac{dz}{d\tau} & \dfrac{dt}{d\tau} \end{bmatrix} = \begin{bmatrix} \gamma & v\gamma \\ v\gamma & \gamma \end{bmatrix} = \begin{bmatrix} \cosh\chi & \sinh\chi \\ \sinh\chi & \cosh\chi \end{bmatrix} \tag{10.41}$$

We have calculated the rotation matrix without having to exponentiate the matrix. Remember, each trigonometric function measures the rate of change of the other trigonometric functions with respect to the angle. What we have above is the measuring of the rate of change of the projection on to the axis of the unit vector as it rotates – this is the trigonometric functions.

Aside: An alternative definition of the exponential function:

Differentiate a C_2 space-time vector with respect to the length of that vector and add the two resulting trigonometric functions to get:

$$\cosh \chi + \sinh \chi = 1 + \frac{\chi^2}{2!} + \ldots + \frac{\chi}{1!} + \frac{\chi^3}{3!} + \ldots = \exp \chi \quad (10.42)$$

Thus, the exponential function (which is the god of all things) comes from the finite group, C_2. It would work, in a more complicated way, with any finite group, actually.

10.4 DIFFERENTIATION OF VECTOR FIELDS AND SCALAR FIELDS

A vector field has two derivatives known as $\{Div, Curl\}$. A scalar field has one derivative known as $\{grad\}$. We show how to derive these in the finite group spaces. In general, when working in the geometric spaces that are derived from the finite groups, we take the differential when we differentiate. Consider the vector field in Euclidean space:

$$F = \begin{bmatrix} u(x,y) \\ v(x,y) \end{bmatrix} \equiv u(x,y) + iv(x,y) = \begin{bmatrix} u(x,y) & v(x,y) \\ -v(x,y) & u(x,y) \end{bmatrix} \quad (10.43)$$

We take the differential as:

$$d\begin{bmatrix} u(x,y) & v(x,y) \\ -v(x,y) & u(x,y) \end{bmatrix} = \frac{\partial \begin{bmatrix} u(x,y) & v(x,y) \\ -v(x,y) & u(x,y) \end{bmatrix}}{\partial \begin{bmatrix} x & y \\ -y & x \end{bmatrix}} \quad (10.44)$$

We cannot differentiate directly with respect to the imaginary part of a complex number, and so we have to do it indirectly. We can differentiate with respect to the real part of a complex number, and so we need to make a few arithmetic adjustments:

$$d\begin{bmatrix} u(x,y) & v(x,y) \\ -v(x,y) & u(x,y) \end{bmatrix} = \cfrac{\dfrac{\partial}{\partial}\begin{bmatrix} u(x,y) & v(x,y) \\ -v(x,y) & u(x,y) \end{bmatrix}}{\dfrac{\partial}{\partial}\begin{bmatrix} x & 0 \\ 0 & x \end{bmatrix}}$$

$$+ \cfrac{1}{\begin{bmatrix} 0 & 1 \\ -1 & 0 \end{bmatrix}} \cfrac{\dfrac{\partial}{\partial}\begin{bmatrix} u(x,y) & v(x,y) \\ -v(x,y) & u(x,y) \end{bmatrix}}{\dfrac{\partial}{\partial}\begin{bmatrix} y & 0 \\ 0 & y \end{bmatrix}}$$

$$= \begin{bmatrix} \dfrac{\partial u}{\partial x} & \dfrac{\partial v}{\partial x} \\ -\dfrac{\partial v}{\partial x} & \dfrac{\partial u}{\partial x} \end{bmatrix} + \begin{bmatrix} 0 & -1 \\ 1 & 0 \end{bmatrix}\begin{bmatrix} \dfrac{\partial u}{\partial y} & \dfrac{\partial v}{\partial y} \\ -\dfrac{\partial v}{\partial y} & \dfrac{\partial u}{\partial y} \end{bmatrix} \quad (10.45)$$

$$= \begin{bmatrix} \dfrac{\partial u}{\partial x} + \dfrac{\partial v}{\partial y} & \dfrac{\partial v}{\partial x} - \dfrac{\partial u}{\partial y} \\ -\left(\dfrac{\partial v}{\partial x} - \dfrac{\partial u}{\partial y}\right) & \dfrac{\partial u}{\partial x} + \dfrac{\partial v}{\partial y} \end{bmatrix}$$

$$= \begin{bmatrix} Div(F) & Curl(F) \\ -Curl(F) & Div(F) \end{bmatrix}$$

The expressions within the matrix are known as the divergence and the curl of the vector field. With a scalar field, we get the gradient of the scalar field:

$$d\begin{bmatrix} u(x,y) & 0 \\ 0 & u(x,y) \end{bmatrix} = \begin{bmatrix} \dfrac{\partial u}{\partial x} & 0 \\ 0 & \dfrac{\partial u}{\partial x} \end{bmatrix} + \begin{bmatrix} 0 & -1 \\ 1 & 0 \end{bmatrix}\begin{bmatrix} \dfrac{\partial u}{\partial y} & 0 \\ 0 & \dfrac{\partial u}{\partial y} \end{bmatrix} \quad (10.46)$$

$$= \begin{bmatrix} \dfrac{\partial u}{\partial x} & -\dfrac{\partial u}{\partial y} \\ \dfrac{\partial u}{\partial y} & \dfrac{\partial u}{\partial x} \end{bmatrix} = Grad(F) \quad (10.47)$$

We will use this method of differentiation in the later chapters of this book, and we will prefer it to the standard way of differentiating with respect to the invariant interval.

If we do the above with space-time, we get:

$$d\begin{bmatrix} u(t,z) & v(t,z) \\ v(t,z) & u(t,z) \end{bmatrix} = \dfrac{\partial\begin{bmatrix} u(t,z) & v(t,z) \\ v(t,z) & u(t,z) \end{bmatrix}}{\partial\begin{bmatrix} t & 0 \\ 0 & t \end{bmatrix}}$$

$$+ \dfrac{1}{\begin{bmatrix} 0 & 1 \\ 1 & 0 \end{bmatrix}}\; \dfrac{\partial\begin{bmatrix} u(t,z) & v(t,z) \\ v(t,z) & u(t,z) \end{bmatrix}}{\partial\begin{bmatrix} z & 0 \\ 0 & z \end{bmatrix}} \qquad (10.48)$$

$$= \begin{bmatrix} \dfrac{\partial u}{\partial t} + \dfrac{\partial v}{\partial z} & \dfrac{\partial v}{\partial t} + \dfrac{\partial u}{\partial z} \\ \dfrac{\partial v}{\partial t} + \dfrac{\partial u}{\partial z} & \dfrac{\partial u}{\partial t} + \dfrac{\partial v}{\partial z} \end{bmatrix}$$

$$= \begin{bmatrix} Div(F) & Curl(F) \\ Curl(F) & Div(F) \end{bmatrix}$$

$$: |Div(F)| > |Curl(F)|$$

Notice that the divergence and curl in space-time are not quite the same as in Euclidean space. In particular it is brought to the reader's notice that the curl in space-time is symmetric in that $\dfrac{\partial v}{\partial t} + \dfrac{\partial u}{\partial z} = \dfrac{\partial u}{\partial z} + \dfrac{\partial v}{\partial t}$ while the curl in Euclidean space is anti-symmet-

ric in that $\dfrac{\partial v}{\partial x} - \dfrac{\partial u}{\partial y} = -\left(\dfrac{\partial u}{\partial y} - \dfrac{\partial v}{\partial x} \right)$. We normally associate the curl of

a vector field with the "force" of that field. Thus, we see that space-time has symmetric forces whereas Euclidean space has anti-symmetric forces. Notice also that the determinant of the above matrix is:

$$\det\left(\begin{bmatrix} \dfrac{\partial u}{\partial t} + \dfrac{\partial v}{\partial z} & \dfrac{\partial v}{\partial t} + \dfrac{\partial u}{\partial z} \\ \dfrac{\partial v}{\partial t} + \dfrac{\partial u}{\partial z} & \dfrac{\partial u}{\partial t} + \dfrac{\partial v}{\partial z} \end{bmatrix} \right) = \left(\dfrac{\partial u}{\partial t} + \dfrac{\partial v}{\partial z} \right)^2 - \left(\dfrac{\partial v}{\partial t} + \dfrac{\partial u}{\partial z} \right)^2 \qquad (10.49)$$

The determinant of a matrix is invariant under change of basis (rotation is one such change of basis). Thus we have that, in space-time:

$$Div^2(F) - Curl^2(F) = \text{constant} > 0 \qquad (10.50)$$

for vector fields over space-time (other types of space have a similar result). For a stationary observer, $\dfrac{\partial u}{\partial z} = \dfrac{\partial v}{\partial z} = 0$. It is only by taking a "God's eye view" that we get the above relationship between the divergence and curl of a vector field in space-time.

10.5 POTENTIALS

A potential is something that we differentiate to get a field. Within electromagnetism, we have two potentials; one is a scalar potential, ϕ, and the other is a vector potential, $\vec{A}$. (We will do more on this later.) We get the electric field, $\vec{E}$, and the magnetic field, $\vec{B}$, by differentiating these potentials as:

$$\vec{B} = curl(\vec{A})$$
$$\vec{E} = -\frac{\partial \vec{A}}{\partial t} - grad(\phi) \qquad (10.51)$$

We get the gravitational field by differentiating the gravitational potential, Φ, as:

$$\vec{a} = -\nabla\Phi \qquad (10.52)$$

Aside: It used to be thought, and to some extent still is, that potentials do not really exist and that only the fields that we get from them actually exist. It was thought that potentials did not produce any observable effects to confirm their existence. The potential was thought of as only a mathematical artifice. However, in 1959, Aharonov (1932–) and Bohm[4] (1917–1992) proposed an experiment that would be capable of detecting the potential. This experiment was done by Chambers[5] in 1960 and showed that potentials do

---

[4]. Y. Aharonov & D.Bohm, Phys. Rev. 115, 485 (1959).
[5]. R.G. Chambers, Phys. Rev. Lett. 5, 3, (1960).

exist and are detectable. The effect is known as the Aharonov-Bohm effect. It compels us to take the view that the potential does exist and has to be thought of as a physical field that really exists and is directly observable[6]. The Aharonov-Bohm effect is concerned with the direction of a unit vector in the 2-dimensional Euclidean complex plane. This direction is known as "the phase" of the vector. It seems that an electromagnetic potential is nothing more than a vector in the Euclidean complex plane, one at each point in space-time, that points in a different direction in the Euclidean complex plane at each different point in space-time. Thus, it seems, we are laying one type of space, the Euclidean complex plane, over another type of space, space-time, and the potential is just how the axes of the two spaces are oriented relative to each other at each point in space-time. The phases form the Lie group $U(1)$. We will meet potentials when we consider electromagnetism.

10.6 FIVE FUNDAMENTAL VECTORS

Within mechanics, there are five fundamental vectors. Although this is true in space-time, and other types of space, the reader will be more familiar with the 3-dimensional spatial vectors together with time in the Newtonian world view. We use this scenario to illustrate the five fundamental vectors. The first is the displacement (position) vector:

$$\vec{r} = [x \quad y \quad z] \tag{10.53}$$

Differentiating this with respect to time gives the velocity vector:

$$\vec{v} = \frac{d\vec{r}}{dt} = \left[\frac{dx}{dt} \quad \frac{dy}{dt} \quad \frac{dt}{dt} \right] \tag{10.54}$$

Differentiating this with respect to time gives the acceleration vector:

$$\vec{a} = \frac{d\vec{v}}{dt} = \left[\frac{d^2x}{dt^2} \quad \frac{d^2y}{dt^2} \quad \frac{d^2t}{dt^2} \right] \tag{10.55}$$

[6.] Interested readers are directed to H. Erlichson, Amer. Jour. Phys, 38, 162 (1970).

Multiplying the velocity vector by the mass gives the momentum vector, and multiplying the acceleration vector by the mass gives the force vector:

$$\vec{p} = m\vec{v}$$
$$\vec{f} = m\vec{a}$$

(10.56)

These five vectors are the basis of mechanics in any space[7]. They are not the only vectors. When we look at electromagnetism we will introduce the electric field vector and the magnetic field vector.

This has been quite a heady chapter. We summarize it. The physically measurable quantities of a vector field are its strength (length of vector) and its direction (angles between vectors). These physically measurable quantities are invariant under change of co-ordinate system, which includes rotation of the observer's point of view. These physical things are calculated by the dot product of one vector with itself and by the dot product of two vectors. The dot product of a vector with itself is called the norm of the algebra.

The main point is that the dot product of a vector field is invariant under rotational transformation. Worth emphasis, methinks!

The dot product (also called the inner product) is invariant under rotation.

Rotation does not change the length of the vector or the angle between two vectors.

Change of velocity is rotation in space-time. If we change the velocity of a vector field (an electric field say by setting an electric charge in motion) the dot product of that vector field (the electromagnetic field) will be unchanged even as the components of the vector field (the electric field and the magnetic field) do change. In mathematics:

$$\det\left(\begin{bmatrix} \gamma & v\gamma \\ v\gamma & \gamma \end{bmatrix}\begin{bmatrix} E & B \\ B & E \end{bmatrix}\right) = \det\left(\begin{bmatrix} \gamma E + v\gamma B & \gamma B + v\gamma E \\ \gamma B + v\gamma E & \gamma E + v\gamma B \end{bmatrix}\right)$$
$$= \det\left(\begin{bmatrix} E & B \\ B & E \end{bmatrix}\right)$$

(10.57)

[7.] Some authors posit six fundamental vectors. They include a vector that is tangent to a moving object, but we have no need of this.

EXERCISES

1. What is the length of the space-time vector $\begin{bmatrix} 5 & 4 \\ 4 & 5 \end{bmatrix}$?

2. What is the length of the space-time vector
$\begin{bmatrix} \cosh \chi & \sinh \chi \\ \sinh \chi & \cosh \chi \end{bmatrix}$?

3. What is the space-time angle $(\cosh(\ ) \text{ equals dot-product})$
between the space-time vectors $\begin{bmatrix} 5 & 4 \\ 4 & 5 \end{bmatrix} \& \begin{bmatrix} 2 & 1 \\ 1 & 2 \end{bmatrix}$?

4. Differentiate the space-time vector field to get the space-time divergence and the space-time curl of the space-time
vector field $\begin{bmatrix} x^2 + 2y & y^3 \\ y^3 & x^2 + 2y \end{bmatrix}$.

5. Differentiate the space-time scalar field to get the space-time gradient of: $\begin{bmatrix} x^3 + 2y^2 & 0 \\ 0 & x^3 + 2y^2 \end{bmatrix}$.

6. What is the cross product of the $\mathbb{R}^3$ vectors: $\{[1\ 2\ 0],$ $[4\ 8\ 0]\}$?

11

THE NATURE OF VELOCITY

Within special relativity, there is not a special reference frame that can consider itself to be absolutely stationary. Two observers who are in motion relative to each other may both take the view that they are the observer at rest and that it is the other observer who is moving. This is the reason for the word "relativity" in the name of the theory of special relativity. The special is that it applies to a "special" space without gravitational distortions.

11.1 ACCELERATION

Contrast this with acceleration. There seems to be an absolute zero acceleration reference frame against which anybody can be compared to see if it is accelerating. It seems that this acceleration reference frame is either absolute space or we must take a Machian view of space. However, this "absolute acceleration" is not like Newton's absolute velocity; each observer has her own absolute acceleration. Two observers in relative motion to each other will disagree about the magnitude of the acceleration of a third observer. It is not the case that the acceleration of a body is measured to be the same against an absolute space for all observers.

That any observer can consider themselves to be stationary (that is moving at zero velocity) corresponds to the ability of any observer to draw their space-time co-ordinate system at an orientation that aligns the time axis with the horizontal. We can arbitrarily choose the "horizontal" time axis to point in any direction we wish, and we usually choose it to point horizontal with respect to ourselves (which makes us stationary). Acceleration is rotating that co-ordinate system (or rotating the whole universe) so that the horizontal axis points in a different direction. Why does space-time rotation require a force proportional to the mass of the body being accelerated? Rotation in Euclidean space does not require a force; once a body is rotating in Euclidean space, it takes no force to keep it rotating.

Euclidean rotation is effectively morphing the cos( ) function into the sin( ) function. These functions are identical except for a 90° displacement (see graphs). Space-time rotation is effectively morphing the cosh( ) function into the sinh( ) function. These functions are not identical. Perhaps this is the reason space-time rotation requires a force.

11.2 A CLOCKWISE ANGLE EQUALS AN COUNTERCLOCKWISE ANGLE

Although observers may disagree over who is at rest and who is moving, they will both agree on the value of the velocity difference between them. That is, they will both agree on the space-time angle between their velocity vectors in space-time. If observer A thinks that observer B is moving at $0.5c$ while he is stationary, then observer B will think that observer A is moving at $0.5c$ while she is stationary.

Why should it be that the two observers agree about their mutual velocity? It is because a "clockwise" rotation through the space-time angle χ is equivalent to an "counterclockwise" rotation through the same space-time angle, χ. Ultimately, this is because the graph of the cosh( ) function is symmetric about the vertical axis (see earlier graphs); or, perhaps, the graph of the cosh( ) function is symmetric about the vertical axis because a clockwise space-time angle is equal to an counterclockwise space-time angle; or, perhaps, it is all down to the form of the exponential function. Let us look at that in terms of rotation matrices. We have:

$$\begin{bmatrix} \cosh\chi & \sinh\chi \\ \sinh\chi & \cosh\chi \end{bmatrix} \begin{bmatrix} \cosh(-\chi) & \sinh(-\chi) \\ \sinh(-\chi) & \cosh(-\chi) \end{bmatrix}$$

$$= \begin{bmatrix} \cosh\chi & \sinh\chi \\ \sinh\chi & \cosh\chi \end{bmatrix} \begin{bmatrix} \cosh\chi & -\sinh\chi \\ -\sinh\chi & \cosh\chi \end{bmatrix} \tag{11.1}$$

$$= \begin{bmatrix} \cosh^2\chi - \sinh^2\chi & 0 \\ 0 & \cosh^2\chi - \sinh^2\chi \end{bmatrix} \tag{11.2}$$

$$= \begin{bmatrix} 1 & 0 \\ 0 & 1 \end{bmatrix}$$

This says that if we rotate through angle χ and then rotate back through the same angle χ, we get back to where we started (the unit matrix is the identity). It seems entirely reasonable; it is entirely reasonable; but it would not happen unless $\cosh(-\chi) = \cosh\chi$ and $\sinh(-\chi) = -\sinh\chi$. The reader should look at the graphs of {cosh( ), sinh( )} and see how they are symmetrical/anti-symmetrical about the vertical axis. This symmetry of the trigonometric functions is called evenness and oddness. The cosh( ) function is an even function, and the sinh( ) function is an odd function.

Let us do the calculation above again, but let us pretend that both the trigonometric functions are odd. We get:

$$\begin{bmatrix} \cosh\chi & \sinh\chi \\ \sinh\chi & \cosh\chi \end{bmatrix} \begin{bmatrix} \cosh(-\chi) & \sinh(-\chi) \\ \sinh(-\chi) & \cosh(-\chi) \end{bmatrix}$$

$$= \begin{bmatrix} \cosh\chi & \sinh\chi \\ \sinh\chi & \cosh\chi \end{bmatrix} \begin{bmatrix} -\cosh\chi & -\sinh\chi \\ -\sinh\chi & -\cosh\chi \end{bmatrix} \tag{11.3}$$

$$= \begin{bmatrix} -\cosh^2\chi - \sinh^2\chi & -2\cosh\chi\sinh\chi \\ -2\cosh\chi\sinh\chi & -\cosh^2\chi - \sinh^2\chi \end{bmatrix}$$

If it were the case that both trigonometric functions were odd, a "clockwise" rotation through χ followed by an "counterclockwise" rotation through χ would not get us back to where we started. The reason that observers agree on the value of their mutual velocity is because the cosh( ) function is an even function, and the sinh( )

function is an odd function. Yet again, thank you to the exponential function.

The symmetry of the cos( ) function can be thought of as the symmetry of the circle. The cos( ) function is the length along the horizontal axis corresponding to a particular angle. Thinking of starting two vectors at the vertical axis and moving one clockwise and the other counterclockwise away from the vertical axis, one can see that the projections on to the horizontal axis are equal for equal angles. This is the even symmetry of the cos( ) function. The even symmetry of the cosh( ) function is the same, but, in the hyperbolic case, it is harder for the reader to envisage.

Aside: The 2-dimensional trigonometric functions are no more than two-way splittings of the exponential function. It is a property of all trigonometric functions in all types of space derived from the finite groups that they have symmetry/anti-symmetry like this oddness/evenness property. However, in higher dimensional spaces derived from the finite groups, this oddness/evenness is not two-way. In 3-dimensional C_3 space, the symmetry/anti-symmetry is three-way and cannot properly be called oddness/evenness. In the 4-dimensional C_4 space, the symmetry/anti-symmetry is four-way. In the 5-dimensional C_5 space, the symmetry/anti-symmetry is five-way, and so on...

In the 4-dimensional $C_2 \times C_2$ space, which will be of great interest to us later, the symmetry/anti-symmetry is also two-way and can properly be called oddness/evenness. This is the case for all spaces derived from finite groups formed by crossing C_2 with itself – $C_2 \times C_2 \times C_2 \times \ldots$.

EXERCISES

1. Assume that both the Euclidean trigonometric functions {cos( ), sin( )} are even and calculate a rotation through θ followed by a rotation through $-\theta$.

12

SIMULTANEITY

Two spatially separated events (crosses) are simultaneous if they happen at the same time; that is, two spatially separated events are simultaneous to an observer if the straight line drawn through them is perpendicular to the time axis of that observer.

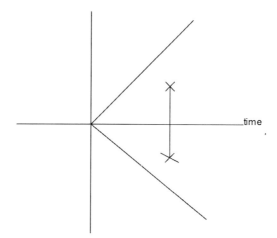

Spatially separated simultaneous events

Note that the line connecting the two events (crosses) is steeper than the 45° lines we have drawn on the diagram.

Of course, observers moving at different velocities will have their time axes oriented in different directions (rotated), and so what is a straight line through two spatially separated events perpendicular to one observer's time axis will not be a straight line that is perpendicular to the time axis of an observer who is moving relative to the first observer.

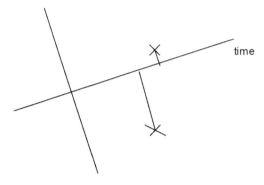

Spatially separated not simultaneous events

Thus, two spatially separated events that are simultaneous for one observer are not simultaneous to any observer who is moving relative to the first observer. That was easy!

12.1 ANOTHER VIEW

A train is moving along a railway track at half the speed of light. At midnight, it passes a stationary observer on the station platform. At that very instant, two bolts of lightning strike the railway track. One strikes exactly two light seconds in front of the train; the other strikes exactly two light seconds behind the train. After two seconds, the flashes from the lightning strikes simultaneously reach the stationary observer. In the stationary observer's view, the lightning strikes were simultaneous. However, by the time the light from the flashes has travelled along the railway line to the stationary observer, the train has moved towards the front flash and is now only one light second away from that front flash but is three light seconds away from the rear flash. The light from the rear flash has further to travel than the light from the front flash, and so it will reach an observer on the train later than the light from the front flash. To the observer on the train, the two lightning flashes are not simultaneous. This explanation dates back to Einstein.

If the two events are not spatio-temporally separated, then the straight line drawn through them is just a dot, and every observer will be able to draw a line to this dot that is perpendicular to their time axis, and so the two coincident events will be simultaneous for every observer.

The two different crosses on the paper represent two events that are separated in space-time. They could be separated in only space, or they could be separated in only time, or they could be separated in a mixture of space and time depending upon how one rotates one's space-time axes. Simultaneity of two events is just a special orientation of the space-time axes that gives a zero amount of time between two events.

12.2 CAUSE AND EFFECT – ORDERED EVENTS

The limiting velocity of space-time is a limit on rotation in space-time. Consider a sheet of paper with the time and space axes drawn upon it at 90° to each other. Draw two small crosses anywhere on the paper to represent two events and draw a line through these two events (crosses).

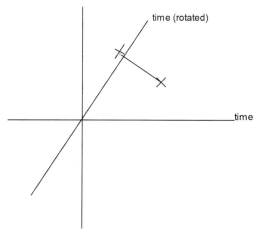

If the rotations in space-time available to us were not restricted, it would be possible to rotate the space-time axes first to a position where the time axis is perpendicular to the straight line through the crosses and then to continue the rotation until the temporal order of the two events (crosses) was reversed (as measured along the time axis). These rotations of axes would be different velocities. So, if the rotations in space-time available to us were not restricted, by putting our foot on the accelerator, we would be able to reverse the time-order of any two spatio-temporally separate events in space-time; if the events influenced each other (one caused the other), then, by accelerating, we would be able to change cause into effect and effect into cause.

It gives one pause, does it not, to think that by changing his velocity your author could make his death precede his birth (leaving no time to write these words). Well, not quite. If your author could travel at the speed of light, then your author's birth and his death would be simultaneous (in the stationary observer's view) because the time interval experienced by a traveller moving at the speed of light is zero (in the stationary observer's view), but your author is restricted from moving at or greater than the speed of light. The rotations (velocities) available to us in space-time are restricted. They are restricted to those velocities between $\pm c$.

Upon a piece of paper draw the space-time axes and two lines at 45° to the horizontal. The lines represent the limiting velocities $\pm c$. They are the asymptotes, on Euclidean paper, of the hyperbola, which is the circle of hyperbolic space (see next chapter). Now place two crosses anywhere in the area between these lines on the positive time side of the paper. Initially, we will place them so that the line connecting them is less steep than the asymptotes (see diagram below).

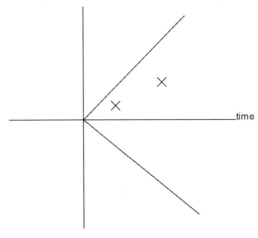

In the diagram, the cross (event) nearest to the origin is the first event because it happens at an earlier time, and the cross furthest from the origin is the later event. The game is to try and reverse the time-order of the events by rotating the axes (leaving the crosses unmoved). You will find that you can alter the time-order of the events only if:

a. you rotate the time axis outside of the 45° lines – that is go to a velocity faster than the speed of light, or if:

b. the line connecting the two crosses is steeper than the asymptotes (which it is in the next diagram).

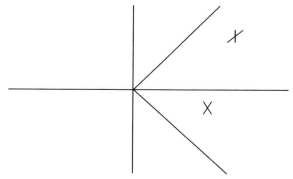

If the line connecting the crosses is steeper than the asymptotes, then the events (crosses) are at such a distance from each other in space and at such proximity to each other in time that they cannot be both visited by anything traveling less than, or equal to, the speed of light – they are effectively in parts of the universe so distant from each other in space and yet so close to each other in time that they can never cause each other because nothing can travel at a speed greater than the speed of light between them. Thus one of them cannot be the cause of the other. Looking back to the first diagram of this chapter, the two simultaneous events are so spatially distant from each other that they cannot influence each other (say by sending a message) in the time (zero) between them.

The mathematics of the limiting velocity is the ratio of the sinh( ) function to the cosh( ) function. It is because this ratio tends to unity as the space-time angle increases that rotation beyond 45° in space-time is impossible. (That's Euclidean degrees not space-time degrees, but you get the idea.) This is why we have cause and effect (ordered events) in the universe. It is remarkable that this ratio is such that we can rotate all the way to the point of making your author's birth simultaneous with his death but not beyond that point to where his death precedes his birth. What would it do to the laws of physics if kettles sometimes boiled before they were filled with water?

Since the time-order of events is due to the nature of the hyperbolic trigonometric functions {cosh( ), sinh( )}, history cannot be reversed because of the nature of those functions. Those functions are splittings of the exponential function – the god of all things.

12.3   BACK TO SIMULTANEITY

Two events cannot be simultaneous for any observers if the line connecting them is less steep than the 45° asymptotes. Such events are just too far separate in time. This means that, within a limited spatial area, some events will happen after other events for all observers. In the extreme, that is for a stationary observer, the spatial area is of zero extent, and so, for a stationary observer, all events will follow in a definite sequence. We are all stationary observers, and we call this sequence of events history.

If, within a limited spatial area, some events happen after, or before, (in time) other events for all observers, then there are events in the universe that everyone agrees are in the past and there are (or will be) events in the universe that no observer has yet seen and are agreed by all observers to be in the future. Thus, within a limited spatial area, the universe has a definite history (in the distant past) and a definite future (in the distant future) and a fuzzy bit around the present. How extensive is the limited spatial area? It is the observable universe. If the universe is 13.8 billion years old, the limited spatial area is a sphere of 13.8 billion light years radius around the observer.

12.4   COMPARISON OF OBSERVATIONS

If two observers want to compare their observations, they must first set their axes in the same orientation or know how to adjust their data to take account of the difference in co-ordinate system orientation. Thus, meaningful comparisons must be done at the same velocity (with the space-time axes in the same orientation) or must be done taking account of the different orientations. The theory of special relativity is the understanding we have of how to take such account of the different orientations of the axes.

If an observer changes the orientation of her axes in space-time (she changes her velocity), then the data that she collected from observing the universe in her first orientation will have to be changed to suit the new orientation of the axes. However, pencil strokes in a notebook do not change when we change velocity. Thus there will seem to be a conflict between the data in the notebook and the data adjusted to the new axes orientation. Such conflicts are known as seeming paradoxes; the most famous of these paradoxes is the twin paradox that we will address later in this book.

The rejection of simultaneity for all observers is the rejection of the Newtonian concepts of absolute space (as far as velocity is concerned) and absolute time for all observers, but this rejection alone would allow the re-ordering of events in any way. The theory of special relativity rejects the Newtonian concepts of absolute space and absolute time but imposes, or rather the exponential function imposes, a limitation upon such re-ordering. Thus, the rejection of absolute space and time is conditional. It is the limiting velocity that leads to the order of events being preserved for all observers. The limiting velocity derives from the nature of the hyperbolic trigonometric functions. Thus, because the order of events is preserved, cause and effect is preserved. A marvellous thing is the exponential function!

WORKED EXAMPLES

1. Relative to a stationary observer with time axis in the horizontal direction, what is the slope of the time axis of an observer moving at half the speed of light?

    Ans: The time axis of the moving observer will intersect the unit invariant interval hyperbola (space-time circle – see next chapter) at a particular point – just like a sloping radius of a Euclidean circle intersects the circle. When we know that point, we can calculate the slope of the moving observer's time axis. The time co-ordinate of that point is given by the Lorentz transformation of the point on the stationary observer's time axis corresponding to $t_0 = 1$:

    $$t' = \frac{t_0}{\sqrt{1 - \frac{\left(\frac{c}{2}\right)^2}{c^2}}} = \frac{2}{\sqrt{3}} \approx 1.15 \tag{12.1}$$

    We get the space co-ordinate from the equation of the hyperbola – which is the invariant interval.

    $$t^2 - z^2 = 1 \Rightarrow z = \sqrt{\frac{4}{3} - 1} = \frac{1}{\sqrt{3}} \tag{12.2}$$

    Thus, the point of intersection of the moving observer's time axis with the hyperbola is $\left(\frac{2}{\sqrt{3}}, \frac{1}{\sqrt{3}}\right)$. The moving observer's time axis is a straight line that passes through the origin. Hence it is described by the function: $z = mt$. This implies $m = \frac{1}{2}$. This is the slope of the moving observer's time axis relative to the horizontal time axis of the stationary observer.

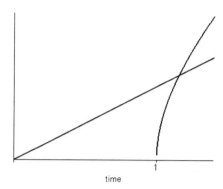

time

1. An observer sees two events that occur at [2, 0.5] and [2, 0] in space-time to be simultaneous. How far apart in time are these two events for an observer moving at half the speed of light relative to the first observer?

Ans: We draw the space-time axes on a sheet of paper with the time axis horizontal and the space axis vertical. We can impose the 45° asymptotes on to the paper if we wish. From A1 above, we have that the observer traveling at half the velocity of light has a time axis that is at 22.5° to the time axis of the stationary observer; we draw this on the paper. We place a dot at each of the given space-time positions. All we have to do is drop lines perpendicular to the moving observer's time axis from each dot and calculate (with simple triangles) the distance along the moving observer's time axis between where the perpendicular lines cut the moving observer's time axis.

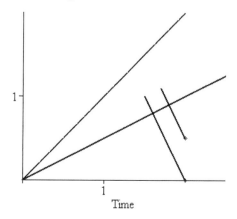

Time

EXERCISES

1. Playing with crosses as above can very much deepen the reader's understanding of why the universe is ordered and with cause and effect. The reader is urged to re-read this chapter and to play with the crosses. Try putting two crosses such that the line joining them is vertical.

2. Taking the units of space to be the same as the units of time, two events occur at [2, 1] & [2.25, 0]. At what velocity need an observer travel to see these events as simultaneous? – It's just simple Euclidean geometry with triangles.

3. a. Calculate a general formula for the slope of a moving observer's time axis relative to a horizontal stationary observer's time axis – enter velocity as a fraction of c.

   b. Calculate a general formula for the line perpendicular to the moving observer's time-axis that intersects a particular given point.

   c. Given the co-ordinates of two events (points in space-time), use the above formulae to calculate the time difference (distance along the moving observer's time axis) between the two events – the Euclidean Pythagoras theorem might be useful.

13

THE LORENTZ TRANSFORMATION

Special relativity is Lorentz invariant physics. It is the physics of a universe that is the same in all directions in space-time – the same at all velocities. We call a change of velocity a Lorentz transformation, and, if the physics is the same at both velocities, we say that the physics is Lorentz invariant (or that the physics is invariant under Lorentz transformations, or, sometimes, that the physics is invariant under the transformations of the Lorentz group – but see later).

13.1 CIRCLES

The circle is the set of points that correspond to the end of a rotating unit vector. In Euclidean space, this is easy to see.

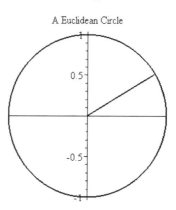

A Euclidean Circle

In space-time, it is not so easy to see this because the circle in space-time appears to be a hyperbola when it is drawn on Euclidean paper – there is a distorting vertical compression in our diagram below.

A Hyperbolic Circle

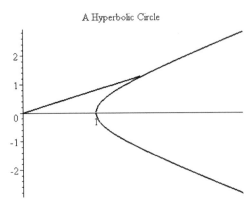

The cosh( ) function is the projection from the hyperbola on to the horizontal axis. It is always greater than one. The sinh( ) function is the projection from the hyperbola on to the vertical axis. The reader might want to look back at the graphs of the hyperbolic trigonometric functions in chapter nine.

In space-time, a rotating vector appears, when drawn on Euclidean paper, to get longer as it rotates. It does get longer if we measure its length using the Euclidean distance function, $d_{euclidean} = \sqrt{x^2 + y^2}$, but this is not the case if we use the distance function of space-time, which is, $d_{space-time} = \sqrt{t^2 - z^2}$. If the reader takes the co-ordinates from the axes of any point on the "hyperbola" drawn above and calculates the distance from the origin to that point using the space-time distance function, the reader will find that the distance from the origin to the "hyperbola" is indeed one (just like a Euclidean circle). The hyperbola is the set of points that are distance one from the origin in space-time – a space-time circle.

Notice how the space-time circle (hyperbola) does not rotate all the way through 360° but is confined to the 45° asymptotes. The asymptotes correspond to the limiting velocity (speed of light).

Consider the distance function of Euclidean space. It is:

$$d_{euclidean} = \sqrt{x^2 + y^2} \qquad (13.1)$$

We set it equal to unity to get the equation of the circle in Euclidean space:

$$x^2 + y^2 = 1 \qquad (13.2)$$

To satisfy this, we must have: $|x| \le 1$, $|y| \le 1$. We see that increasing x will mean decreasing y. Now consider the distance function of space-time. It is:

$$d_{space-time} = \sqrt{t^2 - z^2} \qquad (13.3)$$

We set it equal to unity to get the equation of the circle in space-time (which looks like a hyperbola):

$$t^2 - z^2 = 1 \qquad (13.4)$$

Now, we see that, since $z^2 \ge 0$, the equation of the hyperbolic circle requires $t^2 \ge 1$. $|z|$ can take any value provided $|t|$ takes a "slightly bigger" one - think cosh( ) and sinh( ) - think about just beneath the 45° asymptote. The biggest "slightly" is when $z = 0$; at which point $|t| = 1$. We see that increasing t will mean increasing z.

Rotation in the Euclidean space: as one co-ordinate increases, the other decreases.

Rotation in space-time: as one co-ordinate increases, the other increases.

The reader's attention is drawn to the fact that the determinant of a space-time matrix, which is the distance function of space-time, is always positive because cosh( ) > sinh( ).

13.2 THE LORENTZ TRANSFORMATION

The Lorentz transformation[1] is just a rotation of a vector in space-time. Writing the rotation matrix in terms of the velocity and gamma and the co-ordinates of the end of the (horizontal) vector as $\{t_0, x_0\}$, we have:

[1.] Is it the homogeneity of space that ensures any transformation must be linear (i.e.: matrices); or, is it because space is closed linear transformations (finite groups of matrices) that space is homogeneous?

$$\begin{bmatrix} \cosh\chi & \sinh\chi \\ \sinh\chi & \cosh\chi \end{bmatrix}\begin{bmatrix} t_0 & z_0 \\ z_0 & t_0 \end{bmatrix} = \begin{bmatrix} \dfrac{1}{\sqrt{1-v^2}} & \dfrac{v}{\sqrt{1-v^2}} \\ \dfrac{v}{\sqrt{1-v^2}} & \dfrac{1}{\sqrt{1-v^2}} \end{bmatrix}\begin{bmatrix} t_0 & z_0 \\ z_0 & t_0 \end{bmatrix}$$

$$= \begin{bmatrix} \gamma & v\gamma \\ v\gamma & \gamma \end{bmatrix}\begin{bmatrix} t_0 & z_0 \\ z_0 & t_0 \end{bmatrix} \tag{13.5}$$

$$= \begin{bmatrix} \gamma t_0 + v\gamma z_0 & \gamma z_0 + v\gamma t_0 \\ \gamma z_0 + v\gamma t_0 & \gamma t_0 + v\gamma z_0 \end{bmatrix}$$

$$= \begin{bmatrix} t' & z' \\ z' & t' \end{bmatrix} \quad : \quad |t'| > |z'|$$

Notice the commingling of time and space. This leads to:

$$t' = \gamma\left(t_0 + vz_0\right)$$
$$z' = \gamma\left(z_0 + vt_0\right) \tag{13.6}$$

Where $\{t', x'\}$ are the "new" co-ordinates of the end of the rotated vector. We are rotating from the horizontal (time) axis in an "counterclockwise" direction (it could be "clockwise" if we reversed the velocity) through the space-time angle χ.

13.3 TIME DILATION

The time interval between two events, event A and event B, Δt, is given by $t_A - t_B$. After rotation, it is:

$$t'_A - t'_B = \gamma\left(t_{0A} + vz_{0A}\right) - \gamma\left(t_{0B} + vz_{0B}\right)$$
$$= \gamma\left(t_{0A} + vz_{0A} - t_{0B} - vz_{0B}\right) \tag{13.7}$$

In general, this is:

$$\Delta t' = \gamma\Delta t_0 + \gamma v\Delta z_0$$
$$= \cosh(\chi)\Delta t_0 + \sinh(\chi)\Delta z_0 \tag{13.8}$$

Now, for a stationary observer, the two events take place at the same place in space and so $z_{0A} = z_{0B}$. This gives:

$$t'_A - t'_B = \gamma\left(t_{0A} - t_{0B}\right) \tag{13.9}$$

Which is:

$$\Delta t' = \Delta t$$
$$= \cosh(\ )\Delta \tag{13.10}$$

Note that, in contradistinction to the Euclidean cosine function where we have $\cos\theta \le 1 \ \forall \ \theta$, in space-time, we have $\cosh\chi \ge 1 \ \forall \ \chi$. So:

$$\Delta t' > \Delta t_0 \tag{13.11}$$

The time interval has increased as we have rotated.

This says that the duration of the time interval between two events on a moving spaceship is greater than it would be on a stationary spaceship as seen by the stationary observer. The processes of the universe appear to the stationary observer to go slower on moving spaceships. This is the famous time dilation of special relativity. We could have rotated "counterclockwise" instead of "clockwise" to get:

$$\begin{bmatrix} \cosh\chi & -\sinh\chi \\ -\sinh\chi & \cosh\chi \end{bmatrix}\begin{bmatrix} t_0 & z_0 \\ z_0 & t_0 \end{bmatrix} = \begin{bmatrix} \gamma & -v\gamma \\ -v\gamma & \gamma \end{bmatrix}\begin{bmatrix} t_0 & z_0 \\ z_0 & t_0 \end{bmatrix}$$
$$= \begin{bmatrix} \gamma t_0 - v\gamma z_0 & \gamma z_0 - v\gamma t_0 \\ \gamma z_0 - v\gamma t_0 & \gamma t_0 - v\gamma z_0 \end{bmatrix} \tag{13.12}$$
$$= \begin{bmatrix} t' & z' \\ z' & t' \end{bmatrix} \ : \ |t'| > |z'|$$

Leading to:

$$t' = \gamma(t_0 - vz_0)$$
$$z' = \gamma(z_0 - vt_0) \tag{13.13}$$

The time interval, Δt, is given by $t_A - t_B$. After rotation, it is:

$$t'_A - t'_B = \gamma(t_{0A} - vz_{0A}) - \gamma(t_{0B} - vz_{0B})$$
$$= \gamma(t_{0A} - vz_{0A} - t_{0B} + vz_{0B}) \tag{13.14}$$
$$= \gamma(t_{0A} - t_{0B})$$

Leading to the same result:

$$\Delta t' = \gamma \Delta t_0$$
$$= \cosh(-\chi)\Delta t_0 \tag{13.15}$$

Note that $\cosh(-\chi) = \cosh(\chi)$.

The time dilation formula is normally written as:

$$t' = \frac{t_0}{\sqrt{1 - \dfrac{v^2}{c^2}}} \qquad (13.16)$$

This is a little confusing because it is the time interval that has dilated, and the formula ought properly to be written with $\{\Delta t_0, \Delta t'\}$. Again, we have introduced the c^2 factor to adjust the units to those most often used:

$$\Delta t' = \frac{\Delta t_0}{\sqrt{1 - \dfrac{v^2}{c^2}}} \qquad (13.17)$$

This formula says that, if the stationary observer ages by one second, the stationary observer will see the moving observer age by less than one second.

The moving spaceship is moving through space and time; the stationary observer is moving through only time. However, the rate at which they move through space-time (the space-time length of the vector – the distance from the origin to the hyperbola) is the same for both objects; it is one. The only difference between them is a rotation, and so they have to be the same distance from the origin. Because an increase in the space-co-ordinate, z, requires an increase in the time co-ordinate, t, to keep the space-time length of the vector equal to unity (one), as the moving object moves through space, it must move through more time than the stationary observer. The "more time" is the time dilation. There is more time between two events, and so the processes of the universe seem to go slowly on a moving spaceship. Hyperbolic rotation, which is moving the tip of the vector along the hyperbola, increases both the time co-ordinate and the space co-ordinate. This contrasts with Euclidean rotation in which rotation increases one co-ordinate and decreases the other. We repeat this.

Hyperbolic rotation is such that, as the amount of space travelled through increases, so the amount of time travelled through increases. It must do this to keep the (space-time) length of the vector equal to unity. This is because of the minus sign in the distance function.

The (hyperbolic) rotation has taken a vector in space-time that was wholly time (zero space) and changed it to a vector that is part time and part space. Starting with a vector, whose tip was on the hyperbola, and that was lying along the time axis with only a horizontal component, the rotation has rotated this horizontal vector into a vector, whose tip is still on the hyperbola, but that now has both horizontal and vertical components. The vector is the same length (that is space-time length) after rotation as before rotation because the length of a vector is invariant under rotation – constancy of length is, after all, exactly what rotation is about. Hyperbolic rotation is the vector's tip moving along the hyperbola.

Therefore, it ought not to surprise us that the time component is different in the rotated vector than in the non-rotated vector. The surprise we feel comes from the fact that, as one rotates in space-time, both co-ordinates increase whereas we are accustomed, in Euclidean space, to one co-ordinate decreasing as the other increases during rotation. It's all down to that minus sign in the space-time distance function.

Another way, being careful of the units, to write the time dilation formula is:

$$\Delta t' = \Delta t_0 \cosh \chi \tag{13.18}$$

Where:

$$\chi = \cosh^{-1}\left(\frac{1}{\sqrt{1 - \dfrac{v^2}{c^2}}}\right) \tag{13.19}$$

This is just projecting out the time (horizontal in our diagram) component of a non-horizontal vector (whose tip is on the hyperbola). It compares with:

$$\Delta x' = \Delta x_0 \cos \theta \tag{13.20}$$

in the Euclidean complex plane. That is what trigonometric functions do – project out the components on to an axis. So, the stationary observer moves through space-time in the only time direction – no motion through space, which is why he is stationary. We "drop" the moving observer's time component down on to the stationary observer's time axis by using the cosh( ) trigonometric function, and we get the time component of the moving observer as measured on the time axis of the stationary observer. This is just the same as "dropping" the x-component of a vector on to the x-axis in Euclidean space by using the cos( ) function. The reader should look at the hyperbola above and imagine dropping a vertical projection on to the horizontal time axis – the position it drops to on the time axis is greater than one; this is time dilation. That's probably worth emphasizing:

Time dilation is no more than using the cosh( ) *function to drop the time component of a space-time vector on to the time axis.*

It is perhaps the most famous result of special relativity that the processes of the universe appear to the stationary observer to slow on a moving spaceship. This has given science fiction writers the way to get heroes and heroines of humankind over the great distances between the stars and out into the cosmos before they grow too old to do anything heroic. To the reader who has not before met relativity theory, the idea of the processes of the universe slowing down might seem unfamiliar. If so, it is unfamiliar because our normal everyday experiences involve things moving at only very small velocities relative to ourselves. Einstein once said, "You never understand relativity, you just get used to it". Your author is not sure Einstein was correct on that point. We are a generation that has been familiarized with time dilation by science fiction, both in books and in films. We are aware that the Apollo astronauts who went to the moon in the 1960s and 1970s are three seconds younger than the rest of us because of the time dilation associated with their velocity on the moon trip.

13.4 WE ALL MOVE THROUGH SPACE-TIME AT THE SPEED OF LIGHT

We are less familiar with the idea that even as we stand still, we are moving through space-time. Standing still is moving through time but not through space. When we are moving, we move through both space and time. The rate at which we move through space-time is the "length" of the velocity vector, and this does not change as we begin to move through space – it just rotates. The velocity at which a stationary object travels through space-time is the same as the velocity at which a moving object travels through space-time – which takes a bit of grasping. Ultimately, it is because $\cosh(0) = 1$ that the velocity vector is length one for a stationary observer and thus we move through time as we stand still. Thus, time never stops flowing because the series expansion of the exponential function begins with a one – $\exp(x) = 1 + \dots$.

This takes a bit of grasping mathematically too. The hyperbolic complex numbers (the space-time numbers) are of the form:

$$\begin{bmatrix} h & 0 \\ 0 & h \end{bmatrix} \begin{bmatrix} \cosh \chi & \sinh \chi \\ \sinh \chi & \cosh \chi \end{bmatrix} \tag{13.21}$$

The $\{h\}$ matrix extends or contracts the length of the vector while the rotation matrix just rotates the vector. We, that are humankind, are stuck with $h = 1$. The rate at which we travel through space-time is unchanging as we change our velocity through space. Compare this with the Euclidean complex numbers as represented by the flat sheet of paper before you. We have:

$$\begin{bmatrix} r & 0 \\ 0 & r \end{bmatrix} \begin{bmatrix} \cos \theta & \sin \theta \\ -\sin \theta & \cos \theta \end{bmatrix} \tag{13.22}$$

There are, for humankind, no restrictions on the value of $\{r\}$. We can get anywhere we want on the sheet of paper by a rotation and an extension or contraction of the radial co-ordinate. There are no obvious algebraic reasons for the value of $\{h\}$ being stuck at 1. It is a mystery, but see later, why $h = 1$ for, seemingly, everything in the universe. However, if it were not so, then the concept of the age of the universe being 13.8 billion years would be completely meaningless! Light travels at velocity $h = 1$, and so it seems that $\{h\}$ is an electromagnetic constant of nature.

If we could contract the length of the vector, $\{h\}$, then perhaps we could contract it all the way through zero and out of the other side. This would be traveling backwards in time – just think of the fun one could have messing with the past. The universe would be a much messier place if we were not stuck with $h = 1$.

13.5 THE FEYNMAN CLOCK

Let us approach time dilation from a different direction, for which we thank Richard Feynman. Imagine a train passing through a station at velocity, say $v = 0.5c$. On the station platform is a physics student with a watch. On the train, is a mathematics student with a pair of mirrors facing each other across the train at right angles to the direction of motion of the train. Between the two mirrors, there is a photon of light bouncing back and forth as it is reflected from each mirror on to the other mirror. According to the mathematics student on the moving train, the light travels a distance of one coach width between the mirrors. However, according to the physics student on the platform, the light not only moves across the train carriage but also moves along the platform with the train.

To the mathematics student on the train, who sees herself as the stationary observer, the width of the carriage is given by $c\Delta t_0$. To the physics student on the platform (who, to the math student on the train, appears to be a moving observer), the light traverses a "hypotenuse" distance of $c\Delta t'$ while the train moves through a distance of $v\Delta t'$:

Now the velocity of light, c, being a physical constant, is the same for both students. By (Euclidean) Pythagoras, we have:

$$c^2 \Delta t'^2 = v^2 \Delta t'^2 + c^2 \Delta t_0^2$$

$$\Delta t'^2 \left(c^2 - v^2 \right) = c^2 \Delta t_0^2$$

$$\Delta t' = \Delta t_0 \, \frac{1}{\sqrt{1 - \dfrac{v^2}{c^2}}}$$

(13.23)

This derivation comes from, and depends upon, the constancy of the velocity of light, c which must be assumed.

13.6 TIME DILATION BY DISTANCE FUNCTION

The length of a vector is unchanged by rotation in space-time. That length for a stationary observer is just (the determinant of the C_2 matrix):

$$length_0 = \sqrt{\Delta t_0{}^2 - \frac{\Delta z_0{}^2}{c^2}} \qquad (13.24)$$

Since $\Delta z_0 = 0$, this is just $v\Delta t_0$ for a stationary observer. For a moving observer, that length is:

$$length' = \sqrt{\Delta t'^2 - \frac{\Delta z'^2}{c^2}} \qquad (13.25)$$

These two lengths are the same because the length is unchanged by rotation. Thus:

$$\sqrt{\Delta t'^2 - \frac{\Delta z'^2}{c^2}} = \sqrt{\Delta t_0{}^2 - \frac{\Delta z_0{}^2}{c^2}}$$

$$\Delta t' \sqrt{1 - \frac{\Delta z'^2}{\Delta t'^2 c^2}} = \Delta t_0$$

$$\Delta t' = \frac{\Delta t_0}{\sqrt{1 - \frac{\Delta z'^2}{\Delta t'^2}\frac{1}{c^2}}} \qquad (13.26)$$

$$\Delta t' = \frac{\Delta t_0}{\sqrt{1 - \frac{v^2}{c^2}}}$$

Straight from the group C_2!

If we do this in Euclidean space, starting from the horizontal x-axis ($y_0 = 0$), with the gradient being denoted by g, we get:

$$\sqrt{\Delta x'^2 + \Delta y'^2} = \sqrt{\Delta x_0{}^2 + \Delta y_0{}^2}$$

$$\Delta x' \sqrt{1 + \frac{\Delta y'^2}{\Delta x'^2}} = \Delta x_0 \qquad (13.27)$$

$$\Delta x' = \frac{\Delta x_0}{\sqrt{1 + g^2}}$$

13.7 EXPERIMENTAL EVIDENCE

Time dilation has been experimentally verified on many occasions. The first experimental verification was done by Ives and

Sitwell[2] in 1938. Ives and Sitwell used the to and fro motion of vibrating ions to measure time dilation to within less than five percent[3]. The experiment was done again, slightly differently, in 1985[4] which again verified time dilation. In 1971, Hafele and Keating simply took very accurate caesium clocks around the world on commercial airliners and recorded the time dilation[5]. The Concorde airliner flying at 2,000 kilometers per hour loses 10^{-8} seconds per hour.

Between ten and sixty kilometers above the Earth's surface, cosmic rays striking oxygen and nitrogen atoms produce muons (elementary particles with mass 207 times the electron mass). The muons decay with a half-life of 1.5×10^{-6} seconds. Muons cannot travel faster than the speed of light (300,000 km/sec) and hence cannot travel more than 0.5 km before half of them have decayed. Clearly, not many will travel 50 km before decaying, but many do. In 1941, Rossi and Hall timed muons traveling from the summit to the foot of Mount Washington and found that their half-lives had been dilated in accordance with the predictions of special relativity[6]. In 1975, CERN timed muon decay using muons in a storage ring moving at velocities very near to the speed of light ($\gamma \approx 29$) and again verified the time dilation of special relativity with an accuracy of 2×10^{-3}.

The muon experiments at CERN subjected muons to centripetal accelerations of 10^{18}g. Such acceleration has no effect on the flow of time. However, gravitational fields affect the flow of time. This is an effect that is explained by general relativity not by special relativity. Clocks at the Royal Greenwich Observatory at an altitude of 80 feet above sea level lose 5 microseconds a year compared with clocks at the National Bureau of Standards at Boulder, Colorado at an altitude of 5,400 feet above sea level. Because the Earth's gravitational field at Greenwich is stronger than at Boulder, the clocks at Greenwich run slower – time is dilated by gravity. Time is not dilated by acceleration.

[2.] Ives & Sitwell did not accept special relativity and preferred the aether theory.

[3.] Ives & Sitwell Opt Soc Am 28 215–226 (1938).

[4.] M. Kaivolo et al Phys. Rev. Lett. 54, 255 (1985).

[5.] J.C. Hafele & R. Keating, Science 177,166 (1972).

[6.] Rossi & Hall Phys Rev 59, 223 (1941).

13.8 LENGTH CONTRACTION

Length contraction is not as "obvious" as time dilation. We reproduce the complications below, but, first, we repeat the explanation given earlier in this book. Imagine that light travels at one meter per second; it is only a matter of which units humankind chooses to use. We take the moving world to have velocity 0.9c. If, in the stationary observer's view, the process of light traveling from one end of a meter rod to the other end of the meter rod takes one second in the "stationary world", then the same process will take 2.29 seconds in the "moving world", as seen by the stationary observer, and so the light will travel only $\dfrac{1}{2.29} = 0.436$ moving meter in one stationary second – that is 0.436 moving meters per stationary second. But light always travels at one meter per second. Therefore, the 0.436 meter of moving length must correspond to one meter of stationary length, and so, length in the "moving world" must appear to the stationary observer to be less than it is in the "stationary world" (0.436 meters to one meter at $v = 0.9c$). Time dilation plus constancy of the speed of light equals length contraction.

Length contraction is concerned with the length of a moving rod as seen by a stationary observer. A stationary observer travels through no space. A stationary observer therefore travels through only time. When we calculate the time dilation from the point of view of the stationary observer, we do it by using the Lorentz transformation and setting $z_0 = 0$. The Lorentz transformation is:

$$t' = \gamma\left(t_0 + vz_0\right)$$
$$z' = \gamma\left(z_0 + vt_0\right) \tag{13.28}$$

With $z_0 = 0$, we have the time component:

$$t' = \gamma\, t_0 \tag{13.29}$$

Which is time dilation.

However, if we want to consider the length of a rod we have to make the measurement in such a way that the space-time distance from one end of the rod to the other end of the rod is pure space (no time). To do this, we need to measure from both ends of the rod at the same time. The stationary observer cannot do this because the rear end of the rod passes him a second or so after the front end of the rod passes him. Only an observer moving with the rod can measure its length in pure space. The stationary observer therefore has to accept the measurement of the moving observer.

The two events of reading the tape measure at the two ends of the rod (measuring the length of the rod) are simultaneous in the reference frame of the observer moving with the rod. They are therefore not simultaneous in the reference frame of the stationary observer.

The moving observer will consider herself stationary with respect to the rod. We therefore have to use the Lorentz transformation from the moving observer. The space part of the Lorentz transformation is:

$$z' = \gamma\left(z_0 - vt_0\right) \tag{13.30}$$

when we set $t_0 = 0$, we find that length dilates (just like with time).

$$z' = \gamma\, z_0 \tag{13.31}$$

Just look at the graph of the hyperbola above, as the time co-ordinate increases, so does the space co-ordinate – when cosh( ) increases, so does sinh( ).

This is the view of the moving observer. So, the moving observer sees the length of the rod to be longer (dilated) when it is in the hands of the stationary observer than when it is in her hands, as seen by the moving observer. The stationary observer has to take the moving observer's word that the rod is longer in his hands than in her hands. This means the stationary observer sees the length of the rod shorten (contract) when it moves. The change of viewpoint often leads to confusion, but we have to change the viewpoint (from stationary observer to moving observer) to measure the length of a moving rod in pure space. Let us try it again. We have

$$z' = \gamma\left(z_0 + vt_0\right)$$
$$z_{moving} = \gamma\left(z_{stationary} + vt_{stationary}\right) \tag{13.32}$$

But we need to swap the point of view from the stationary observer to the moving observer. Thus we swap:

$$z_{stationary} = \gamma \left(z_{moving} - vt_{moving} \right)$$
$$z_0 = \gamma (z' - vt')$$

(13.33)

With $t' = 0$:

$$z_0 = \gamma z'$$

(13.34)

Leading to:

$$z' = z_0 \frac{1}{\gamma}$$

(13.35)

This is usually presented as:

$$\Delta l' = \Delta l_0 \sqrt{1 - \frac{v^2}{c^2}}$$

(13.36)

This is the length contraction formula. This is how the stationary observer sees the length of a moving rod. For comparison, the time dilation formula is:

$$\Delta t' = \frac{\Delta t_0}{\sqrt{1 - \frac{v^2}{c^2}}}$$

(13.37)

When it comes to measuring length, the stationary observer has to take the word of the moving observer, and she says that the stationary meter rod is longer than her moving rod. The stationary observer has to accept the moving observer's measurement, and so the moving meter rod is shorter for the stationary observer than the stationary meter rod.

If the above presentation seems contorted, just think of the stationary observer's view of length contraction as "space has dilated but the rod has stayed the same".

We will eventually be working with 4-dimensional space-time. The time of a moving observer is dilated regardless of the direction in "3-dimensional" space of that observer's motion. The length of a moving observer contracts only in the spatial direction of the motion. Thus, a sphere becomes a flattened sphere when moving as it contracts in only one of its spatial dimensions.

13.9   EXPERIMENTAL EVIDENCE

It is believed, but has never been verified, that the Concorde airliner flying at 2,000 kilometers per hour shortens by 10^{-10} meters. There is no experimental evidence of length contraction of macroscopic objects, but the charge density of a current carrying wire changes appropriately. Further, a moving observer sees a stronger electromagnetic coulomb force between two separated electrically charged objects than is seen by a stationary observer.

Imagine a flat circular disc. When stationary, the circumference of this disc is $circum < \pi D$. When the disc is rotating, the length of the outer edge will contract due to length contraction of moving bodies. Thus, the circumference of a rotating disc is predicted by the theory of special relativity to be less than that of a stationary one, $circum < \pi D$. Although they did not measure length contraction, in 1960, Hay, Schiffer, Cranshaw, and Engelstaff, using Mossbauer resonance[7], verified that there is time dilation at the edge of a rapidly spinning rotor compared to the center of the rotor. Mossbauer apparati can measure the Doppler shift of light from a source moving at only $10^{-7} ms^{-1}$.

13.10 GETTING TECHNICAL

The Lorentz transformation is a linear transformation (it's done with matrices). Because the Lorentz transformation is linear, it maps straight lines to straight lines. This means that a moving object will be seen to move in a straight line through space-time by both a moving observer and a stationary observer. If the Lorentz transformation were not linear, one of the observers would see the object travel in a curved line in space-time and would deduce that it was subject to a force. There would be forces popping into existence everywhere. We could create them by changing our velocity. These forces would correspond to energy gradients, and so the conservation of energy would go out of the window. If we are to avoid such complications,

---

[7.] Hay, Schiffer, Cranshaw & Engelstaff. Phys Rev Lets 4,165 (1960).

empty spaces must be linear. It is remarkable that the universe does things just right! Without linearity, no conservation of energy (and much else goes wrong besides).

13.11 A FEW USEFUL IDENTITIES

For future reference, we end this chapter by appending some useful identities. We first remind the reader that, with the speed of light, $c = 1$:

$$\frac{dt}{d\tau} = \gamma = \frac{1}{\sqrt{1-v^2}} = \cosh \chi \qquad (13.38)$$

$$\frac{dz}{d\tau} = v\gamma = \frac{v}{\sqrt{1-v^2}} = \sinh \chi \qquad (13.39)$$

Now:

$$\gamma^2 v^2 = \frac{v^2}{1-v^2} + 1 - 1 = \frac{v^2}{1-v^2} + \frac{1-v^2}{1-v^2} - 1 = \frac{1}{1-v^2} - 1 = \gamma^2 - 1 \quad (13.40)$$

Which is:

$$\cosh^2 \chi \tanh^2 \chi = \cosh^2 \chi \frac{\sinh^2 \chi}{\cosh^2 \chi} = \sinh^2 \chi = \cosh^2 \chi - 1 \qquad (13.41)$$

And:

$$\gamma^2 v^2 + 1 = \frac{v^2}{1-v^2} + 1 = \frac{v^2 + 1 - v^2}{1-v^2} = \frac{1}{1-v^2} = \gamma^2 \qquad (13.42)$$

Which is:

$$\sinh^2 \chi + 1 = \cosh^2 \chi \qquad (13.43)$$

And:

$$\frac{d\gamma}{dt} = \frac{d}{dt}\left(1-v^2\right)^{-\frac{1}{2}} = -\frac{1}{2}\left(1-v^2\right)^{-\frac{3}{2}}\left(-2v\frac{dv}{dt}\right) = \gamma^3 v\frac{dv}{dt} \quad (13.44)$$

Similarly:

$$\frac{d\gamma}{dz} = \gamma^3 v\frac{dv}{dz} \qquad (13.45)$$

And:

$$\frac{d(v\gamma)}{dt} = \frac{d}{dt}\left(v\left(1-v^2\right)^{-\frac{1}{2}}\right)$$

$$= -\frac{1}{2}\left(1-v^2\right)^{-\frac{3}{2}}\left(-2v\frac{dv}{dt}\right)v + \left(1-v^2\right)^{-\frac{1}{2}}\frac{dv}{dt}$$

$$= \gamma^3 v^2\frac{dv}{dt} + \gamma\frac{dv}{dt}$$ (13.46)

$$= \gamma\left(\gamma^2-1\right)\frac{dv}{dt} + \gamma\frac{dv}{dt}$$

$$= \gamma^3\frac{dv}{dt}$$

Similarly:

$$\frac{d(\gamma v)}{dz} = \gamma^3 v^2\frac{dv}{dz} + \gamma\frac{dv}{dz}$$ (13.46)

$$= \gamma^3\frac{dv}{dz}$$

WORKED EXAMPLES

1. To what velocity does the space-time point $(4, 3)$ correspond? To what space-time angle does this correspond?

    Ans: We have:

    $$\begin{bmatrix} \gamma & v\gamma \\ v\gamma & \gamma \end{bmatrix}\begin{bmatrix} 1 & 0 \\ 0 & 1 \end{bmatrix} = \begin{bmatrix} 4 & 3 \\ 3 & 4 \end{bmatrix} \Rightarrow \gamma = 4 \ \& \ v\gamma = 3 \Rightarrow v = \frac{3}{4}c \ (13.47)$$

    We have:

    $$\begin{bmatrix} \cosh\chi & \sinh\chi \\ \sinh\chi & \cosh\chi \end{bmatrix} = \begin{bmatrix} 4 & 3 \\ 3 & 4 \end{bmatrix} \Rightarrow \tanh\chi = \frac{3}{4} \Rightarrow \chi = 0.9729... \ (13.48)$$

EXERCISES

1. a. What is the distance of the space-time point $\left(\frac{5}{3}, \frac{4}{3}\right)$ from the origin?

      b. What is the distance of the Euclidean space point $\left(\dfrac{5}{3}, \dfrac{4}{3}\right)$ from the origin?

2. What is the time dilation associated with the space-time point $(5, 4)$?

14

VELOCITY AND ACCELERATION TRANSFORMATIONS

Imagine a math student stood upon a (stationary) railway platform. A train carrying a physics student passes the math student at velocity, $v_{physics} = 0.9c \equiv \psi$ - nine tenths of the speed of light where ψ is the corresponding angle in space-time. At the same time, a train carrying a biology student passes the physics student's train with a relative velocity of $v_{bio} = 0.8c \equiv \phi$ - eight tenths of the speed of light where ϕ is the corresponding angle in space-time. Now, the physics student agrees with the biology student that they pass each other at the relative velocity of $v = 0.8c$, and the math student and the physics student agree that they pass each other at the relative velocity of $v = 0.9c$. It is immediately apparent that the biology student in the fastest train passes the math student on the platform at $0.9c + 0.8c = 1.7c$, but this cannot be correct because no-one, not even biology students, can move faster than the speed of light. Let us do the math.

The physics student is rotated in space-time by ψ relative to the math student and the biology student is rotated by ϕ in space-time relative to the physics student. Thus, the biology student is rotated twice in space-time relative to the math student. This is:

$$\begin{bmatrix} \cosh\phi & \sinh\phi \\ \sinh\phi & \cosh\phi \end{bmatrix} \begin{bmatrix} \cosh\psi & \sinh\psi \\ \sinh\psi & \cosh\psi \end{bmatrix}$$

$$= \begin{bmatrix} \cosh\phi\cosh\psi + \sinh\phi\sinh\psi & \cosh\phi\sinh\psi + \sinh\phi\cosh\psi \\ \cosh\phi\sinh\psi + \sinh\phi\cosh\psi & \cosh\phi\cosh\psi + \sinh\phi\sinh\psi \end{bmatrix} \quad (14.1)$$

$$= \begin{bmatrix} \cosh(\phi+\psi) & \sinh(\phi+\psi) \\ \sinh(\phi+\psi) & \cosh(\phi+\psi) \end{bmatrix}$$

cosh( ) is the projection of a moving observer on to the time axis.
sinh( ) is the projection of a moving observer on to the space axis.
The velocity is given by:

$$v = \frac{space}{time} = \frac{\sinh(\ )}{\cosh(\ )} = \tanh(\ ) \quad (14.2)$$

Now, the velocity at which the biology student passes the math student is given by the tanh( ) of the angle $(\phi + \psi)$; that is:

$$\tanh(\phi+\psi) = \frac{\tanh\phi + \tanh\psi}{1 + \tanh\phi\tanh\psi}$$

$$(14.3)$$

$$v_{maths/bio} = \frac{v_{bio} + v_{physics}}{1 + v_{bio}v_{physics}}$$

Adjusting the units of space-time:

$$\tanh(\phi+\psi) = \frac{v_{bio} + v_{physics}}{1 + \dfrac{v_{bio}v_{physics}}{c^2}} \quad (14.4)$$

Putting the numbers in:

$$v_{bio/maths} = \frac{0.8c + 0.9c}{1 + \dfrac{0.8.0.9c^2}{c^2}} = \frac{1.7}{1.72}c < c \quad (14.5)$$

This is called the velocity transformation and is usually presented as:

$$v' = \frac{u_1 + u_2}{1 + \dfrac{u_1 u_2}{c^2}} \quad (14.6)$$

Since we have a limiting velocity in space-time, we cannot have simple addition of velocities. Without a limiting velocity, the events of the universe would not be ordered, and if we were able to simply add velocities, everything would fall to pieces!

14.1 ACCELERATION TRANSFORMATION

Within the literature, there is no preferred way to calculate the acceleration transformation. The acceleration transformation is far less well known than the velocity transformation, and many texts on special relativity do not consider it. This is because it is both a more complicated and a more difficult to understand transformation than most other transformations. We derive a simple version (but not the whole story) of it here in two ways. In a later chapter, we derive a "fuller" version of it using the mathematics of 4-vectors and, after that, an "even fuller" version of it using matrices.

14.2 FIRST DERIVATION OF THE ACCELERATION TRANSFORMATION

Acceleration in the direction of the velocity

Consider a train moving at (say) half the speed of light passing by a stationary platform and accelerating in the direction of its velocity as it passes. Acceleration is meter $\sec^{-2}$. This is length divided by time squared. We have:

$$a' = \frac{l'}{t't'} = \frac{\frac{1}{\gamma}l_0}{\gamma t_0 \gamma t_0} = \frac{1}{\gamma^3}\frac{l_0}{t_0 t_0} = \frac{a_0}{\gamma^3} \qquad (14.7)$$

$$a' = \frac{a_0}{\gamma^3}$$

Where a_0 is the acceleration as is measured by the observer moving with the train and a' is the acceleration measured by the stationary observer on the platform. Note that a_0 is the acceleration that would be measured by the platform-bound observer if the train was (momentarily) stationary on the platform as it accelerated.

So, if a train accelerates at $a_0 = 10$ meters per $\sec^{-2}$ when it is stationary, it will be felt to accelerate at 10 meters per $\sec^{-2}$ by the moving observer aboard the moving train when it is moving at (say) half the speed of light. When the train passes the platform, the stationary observer on the platform will see the train accelerating in the direction of its velocity, relative to herself, not at 10 meters per sec^{-2} but at (when $v = \dfrac{c}{2}$):

$$\frac{10}{\gamma^3} = 10\left(\sqrt{1-\frac{1}{4}}\right)^3 = 10.\frac{3\sqrt{3}}{8} = 6.5 \text{ meter sec}^{-2} \qquad (14.8)$$

Thus, as the train approaches the speed of light relative to the platform, the platform-bound observer will see its acceleration approach zero whereas the train-bound observer will think the train is still accelerating at 10 meters per sec^{-2}.

Acceleration perpendicular to the direction of the velocity

Now let us imagine that the train is accelerating not in the direction of its velocity but in a direction perpendicular to that velocity – think acceleration of a planet moving around a star in a perfectly circular orbit. Because the acceleration is in a direction at right angles to the velocity, there will be no length contraction in that direction, but there will still be time dilation. The above calculation becomes: Acceleration is meter sec^{-2}. This is (non-contracted) length divided by time squared. We have:

$$a' = \frac{l'}{t't'} = \frac{l_0}{\gamma t_0 \gamma t_0} = \frac{1}{\gamma^2}\frac{l_0}{t_0 t_0} = \frac{a_0}{\gamma^2} \qquad (14.9)$$

$$a' = \frac{a_0}{\gamma^2}$$

There is a factor of gamma difference.

14.3   SECOND DERIVATION OF THE ACCELERATION TRANSFORMATION

Using $t' = \gamma\, t_0$:

$$a_0 = \frac{d}{dt_0}(v_0) = \frac{d}{dt'}(v\gamma) = \frac{d}{dt'}\left(v(1-v^2)^{-\frac{1}{2}}\right)$$

$$= -\frac{1}{2}(1-v^2)^{-\frac{3}{2}}\left(-2v\frac{dv}{dt'}\right)v + (1-v^2)^{-\frac{1}{2}}\frac{dv}{dt'} \qquad (14.10)$$

$$= \gamma^3 v^2 \frac{dv}{dt'} + \gamma\frac{dv}{dt'}$$

$$= \gamma(\gamma^2 - 1)\frac{dv}{dt'} + \gamma\frac{dv}{dt'} = \gamma^3\frac{dv}{dt'}$$

In short:

$$a_0 = \frac{d(v\gamma)}{dt'} = \gamma^3 \frac{dv}{dt'} = \gamma^3 a' \qquad (14.11)$$

We will see a third derivation of the acceleration transformation when we study 4-vectors and a fourth one when we use matrices to do what we did with 4-vectors.

We do not need mass to increase towards infinity to prevent us from exceeding the speed of light. If there was no mass, we would still be unable to accelerate through the light barrier (as seen by a stationary observer outside the rocket). The limiting velocity is a feature of space-time; it is nothing to do with mass. It has everything to do with the fact that $\sinh \chi \to \cosh \chi$ as $\chi \to \infty$, and so the ratio of these two functions, which is velocity, tends to unity as the space-time angle, χ, tends to infinity.

There is another complication to acceleration. Imagine a rod moving and accelerating in the direction of its length. As its velocity increases, the rod's length shortens. This means that, if the front end of the rod is accelerating at a_{front}, then either the rear end of the rod must accelerate faster than the front end, $a_{front} < a_{rear}$, or the rod must stretch within itself. This stretching is the same as stretching a stationary rod and is nothing to do with the nature of space-time. This stretching is not the "spagettiffication" of a rod falling into a black hole due to the different strengths of the gravitational field at the ends of the rod. This is a separate effect over and above the "spagettiffication" effect.

EXERCISES

1. A distant green galaxy is receding directly from Earth at a velocity of $0.7c$. A green alien in that galaxy, looking in the same direction as earth-bound observer, sees a distant pink galaxy receding from her at a velocity of $0.9c$. At what velocity does the earth-bound observer see the pink galaxy receding from the Earth?

2. A distant purple galaxy is receding from the Earth at 0.9c. As it recedes, it appears to the earth-bound observer to accelerate at 0.5c per sec. At what rate does a purple alien in the distant galaxy think the purple galaxy is accelerating? Would such acceleration violate the limiting velocity of the universe?

15

THE NATURE OF STRAIGHT LINES AND THE TWIN PARADOX

15.1 THE CALCULUS OF VARIATIONS

The calculus of variations is central to Lagrangian mechanics and particle physics. It is an advanced mathematical topic that has only little (and relatively simple) application in special relativity. We do not here attempt to familiarize the reader with the whole of the calculus of variations; that would be a textbook on its own. The reader might be unfamiliar with the calculus of variations and might struggle to understand the next few pages, but the details are not important. It is the results that are important for our purposes. We will show, by using the calculus of variations, that a straight line is the shortest distance between two points in Euclidean space. This is indeed how Euclid defined the straight line circa 300 BC. We will then show that a straight line is the longest distance between two points in space-time. This means that, in space-time, bent lines are shorter than straight lines. Let me emphasize that:

Euclidean space: A straight line is the *shortest* distance between two points.

Space-time: A straight line is the **longest** distance between two points.

Perhaps you should read that bit again; for someone who has never before met with this fact, it is a bit of a shocker!

For the next few paragraphs, for pedagogical ease, we will work in the 2-dimensional Euclidean space known as $\mathbb{R}^2$. The reader is invited to visualise this space as a flat sheet of paper with a vertical y-axis and a horizontal x-axis. In this space, we have functions of the form $y = f(x)$. We will be seeking the particular function, $f(x)$, that minimises the distance between two points in this space. We already know the answer – the minimum distance between two points is a straight line. A straight line is expressed as the function:

$$y = mx + c \qquad (15.1)$$

We seek to prove that this function is the minimum distance between two points in $\mathbb{R}^2$. We do this with the calculus of variations. It is assumed that the reader is familiar with calculus of normal variables up to and including the ability to calculate the maxima and minima of a function. We begin:

The distance function of the Euclidean plane is:

$$S^2 = x^2 + y^2 \qquad (15.2)$$

We write this in terms of differentials:

$$dS = \sqrt{dx^2 + dy^2} \qquad (15.3)$$

A little manipulation gives us:

$$\frac{dS}{dx} = \sqrt{1 + \left(\frac{dy}{dx}\right)^2}$$

$$\qquad (15.4)$$

$$S = \int dx \sqrt{1 + \left(\frac{dy}{dx}\right)^2}$$

We assume that we do not know what the function $y(x)$ actually is (except that, in this case, we do because we know the answer). The S is considered to be a functional of the unknown function $y(x)$ and is written:

$$S[y(x)] = \int_a^b dx \sqrt{1 + y'(x)^2} \qquad (15.5)$$

By which is meant that $y'(x) = \dfrac{dy}{dx}(x)$ is a particular, but not yet determined function of the variable x. $S[y(x)]$ is the distance between the points $a\&b$ in the Euclidean plane. The idea is that we are going to vary (change) the function $y'(x)$ a little bit instead of varying (changing) the variable, x, a little bit. In normal calculus, we use little bits of variable, $\{dy, dx...\}$. In variational calculus, we use little bits of function $\{\delta(y'(x))\}$. We are going to find the stationary points (that is the maxima and minima) of $S[y(x)]$ in a way very similar to how we find the stationary points of a function in normal calculus.

We are going to vary the function $y(x)$ by adding to it another function, $h(x)$, multiplied by a real variable, ε. We will then let the function $y(x) + \varepsilon h(x)$ approach $y(x)$ by letting $\varepsilon \to 0$ in a way similar to how we let $\delta\varepsilon \to 0$ in normal calculus.

$y(x) + \varepsilon h(x)$ is a slightly different function from $y(x)$, and so $y(x) + \varepsilon h(x)$ represents a slightly different path between the points $a \& b$ from the path represented by $y(x)$. Since all paths go through the points $a \& b$, we have $h(a) = h(b) = 0$.

Now the difference in the lengths of the paths described by $y \& y + \varepsilon h$ is:

$$\delta S = S\big[y(x) + \varepsilon h(x)\big] - S\big[y(x)\big] \tag{15.6}$$

And:

$$S\big[y(x)\big] = \int_a^b dx\sqrt{1 + y'(x)^2}$$

$$\tag{15.7}$$

$$S\big[y(x) + \varepsilon h(x)\big] = \int_a^b dx\sqrt{1 + \big[y'(x) + \varepsilon h'(x)\big]^2}$$

Taylor series expansion in powers of ε gives:

$$S\big[y + \varepsilon h\big] = \int_a^b dx\left(\sqrt{1 + y'^2} + \varepsilon\left(\frac{d}{d\varepsilon}\sqrt{1 + (y' + \varepsilon h')^2}\right)... + O(\varepsilon^2)\right)$$

$$= \int_a^b dx\left(\sqrt{1 + y'^2} + \varepsilon\frac{y'h'}{\sqrt{1 + y'^2}}... + O(\varepsilon^2)\right)$$

$$\tag{15.8}$$

And so:

$$\delta S = S[y + \varepsilon h] - S[y]$$

$$= \int_a^b dx \left(\sqrt{1 + y'^2} + \varepsilon \frac{y'h'}{\sqrt{1 + y'^2}} ... + O(\varepsilon^2) \right) - \int_a^b dx \sqrt{1 + y'(x)^2} \qquad (15.9)$$

$$= \int_a^b dx \left(\varepsilon \frac{y'h'}{\sqrt{1 + y'^2}} ... + O(\varepsilon^2) \right)$$

Since when $S[y]$ is a stationary point, $\delta S = 0$, and $\varepsilon \neq 0$, it follows that:

$$\int_a^b dx \left(\frac{y'}{\sqrt{1 + y'^2}} \right) h' = 0 \qquad (15.10)$$

and here we have the function that is stationary (minimum or maximum). Now, given that $h(a) = h(b) = 0$, the integral between $\{a, b\}$ will be zero if:

$$\frac{y'}{\sqrt{1 + y'^2}} = \text{constant} \qquad (15.11)$$

Thus we have $y'(x) = \text{constant} = m$. Integration gives:

$$S[y] = \int_a^b dx(m) = mx + c \qquad (15.12)$$

This is the equation of a straight line, and so the stationary path is a straight line. We have thus shown that the stationary (maximum or minimum) distance between two points in Euclidean space is given by the function $y = mx + c$ – a straight line. We have not yet shown this straight line to be the minimum (it could be a maximum) distance. The reader should recall that, in normal calculus, the sign of the second differential, $\frac{d^2y}{dx^2}$, determines whether or not we have a maximum or a minimum. Above, when we Taylor expanded $\sqrt{1 + [y'(x) + \varepsilon h'(x)]^2}$, we ignored the ε^2 term, but this is exactly the term we need to determine the nature of the stationary function (max or min?). We have:

$$\sqrt{1 + [y'(x) + \varepsilon h'(x)]^2}$$

$$= \sqrt{1 + y'^2} + \varepsilon \frac{y'h'}{\sqrt{1 + y'^2}} + \varepsilon^2 \frac{h'^2}{2(1 + y'^2)^{\frac{3}{2}}} ... + O(\varepsilon^3) \qquad (15.13)$$

We have, from above that $y'(x) =$ constant. The ε^2 term is therefore positive for all $h(x)$, and we have a minimum. Thus, in Euclidean space, a straight line is the shortest distance between two points.

We now do all of the above with the distance function of space-time. The distance function of the space-time is:

$$S^2 = t^2 - z^2 \qquad (15.14)$$

A little manipulation gives us:

$$S = \int dt \sqrt{1 - \left(\frac{dz}{dt}\right)^2} \qquad (15.15)$$

Which leads to:

$$\frac{z'}{\sqrt{1 - z'^2}} = \text{constant} \qquad (15.16)$$

Integration gives:

$$S[z] = \int_a^b dt(\text{m}) = mt + c \qquad (15.17)$$

Which again is the equation of a straight line, and so, just like in Euclidean space, the stationary path in space-time is a straight line. It is actually the path that an observer moving at constant velocity (could be zero) follows through space-time.

However, the ε^2 term is:

$$-\varepsilon^2 \frac{h'^2}{2\left(1 - z'^2\right)^{3/2}} \qquad (15.18)$$

z' is constant, and the ε^2 term is negative, and we have a maximum. In space-time, a straight line is the longest distance between two points. More re-emphasis:

Euclidean space: A straight line is the **shortest** distance between two points.

Space-time: A straight line is the **longest** distance between two points.

This seems weird, but that is, not very simply, our prejudice. Why should a straight line be the shortest distance between two points?

Why should a straight line not be the longest distance between two points[1]? Now forget $\mathbb{R}^2$; everything we have done applies to the complex plane and to the hyperbolic complex space-time. Euclidean space and space-time both come from the finite group C_2. They are like brother and sister. Furthermore, they are the only spaces that come from this finite group. There is a kind of "completeness" about these two spaces having the two possible but opposite kinds of straight line.

15.2 ANOTHER VIEW

We have, in Euclidean space, for two vectors at angle θ to each other:

$$\left(\vec{V_1} + \vec{V_2}\right) \bullet \left(\vec{V_1} + \vec{V_2}\right) = \vec{V_1} \bullet \vec{V_1} + \vec{V_2} \bullet \vec{V_2} + 2\vec{V_1} \bullet \vec{V_2}$$

$$\left|\vec{V_1} + \vec{V_2}\right|^2 = \left|\vec{V_1}\right|^2 + \left|\vec{V_2}\right|^2 + 2\left|\vec{V_1}\right|\left|\vec{V_2}\right|\cos\theta \qquad (15.19)$$

$$\left|\vec{V_1} + \vec{V_2}\right|^2 \le \left|\vec{V_1}\right|^2 + \left|\vec{V_2}\right|^2 + 2\left|\vec{V_1}\right|\left|\vec{V_2}\right|$$

$$= \left(\left|\vec{V_1}\right| + \left|\vec{V_2}\right|\right)^2$$

The vector $\vec{V_1} + \vec{V_2}$ is a straight line between two points. In Euclidean space, the vector $\vec{V_1} + \vec{V_2}$ has length $\left|\vec{V_1} + \vec{V_2}\right|$ and is shorter than $\left|\vec{V_1}\right| + \left|\vec{V_2}\right|$ (think triangle) because $\cos\theta \le 1$. In space-time, for two vectors at angle χ to each other we have:

$$\left(\vec{V_1} + \vec{V_2}\right) \bullet \left(\vec{V_1} + \vec{V_2}\right) = \vec{V_1} \bullet \vec{V_1} + \vec{V_2} \bullet \vec{V_2} + 2\vec{V_1} \bullet \vec{V_2}$$

$$\left|\vec{V_1} + \vec{V_2}\right|^2 = \left|\vec{V_1}\right|^2 + \left|\vec{V_2}\right|^2 + 2\left|\vec{V_1}\right|\left|\vec{V_2}\right|\cosh\chi \qquad (15.20)$$

$$\left|\vec{V_1} + \vec{V_2}\right|^2 \ge \left|\vec{V_1}\right|^2 + \left|\vec{V_2}\right|^2 + 2\left|\vec{V_1}\right|\left|\vec{V_2}\right|$$

$$= \left(\left|\vec{V_1}\right| + \left|\vec{V_2}\right|\right)^2$$

In space-time, the vector $\vec{V_1} + \vec{V_2}$, which is a straight line between two points, has length $\left|\vec{V_1} + \vec{V_2}\right|$ and is longer than $\left|\vec{V_1}\right| + \left|\vec{V_2}\right|$ because $\cosh\chi \ge 1$ - see graphs.

---

[1] Yes, it does twist the brain a little.

15.3   YET ANOTHER VIEW

In Euclidean space, we have the {3, 4, 5} triangle given by the Pythagoras triple:

$$5^2 = 3^2 + 4^2 \tag{15.21}$$

In space-time, we have the {3, 4, 5} triangle given by the Pythagoras like triple:

$$3^2 = 5^2 - 4^2 \tag{15.22}$$

In both cases the length of the hypotenuse of the triangle is the number that stands alone on the left of the equals sign. We see that in Euclidean space the hypotenuse of a triangle is shorter than the sum of the other two sides: $5 < 3 + 4$. In space-time, we have that the hypotenuse is longer than the sum of the other two sides: $3 > 5 - 4$ - see where the minus sign comes in.

15.4   THE TWIN PARADOX

There is, and has been for decades, a mythology based primarily on what seems to be a contradiction within special relativity. That seeming contradiction is known as the twin paradox. The idea is based upon the fact that velocity is relative and that each of two observers in relative motion may declare themselves to be the stationary observer and declare that the other observer is the moving observer. Since moving clocks run slowly, the moving observer will age more slowly than the stationary observer and, thus, after, say, fifty years, will look appreciably younger than the stationary observer. The paradox is that both observers can equally consider themselves to be moving, and so each will look appreciably younger than the other.

In the case that the two observers keep moving relative to each other rather than one of the observers change their velocity to match the other observer, then they are indeed both younger than the other as seen by the other. It is the change of velocity to match the other that breaks the symmetry of the system. A change of velocity by observer A to match the velocity of observer B is "dropping" the space-time vector of A on to the time axis of observer B. A change of velocity by observer B to match the velocity of observer A is "dropping" the space-time vector of B on to the time axis of observer A. If the velocities never change, then there is no "dropping" on to the other observer's time axis. Which is the younger depends on the point of view (the axes from which you choose to observe).

The twin paradox is usually formulated with two identical twins as the observers. One of the twins stays at home on Earth while the other twin boards a spaceship and travels, at almost light speed, to a nearby star before turning around and coming back to the Earth. From the Earth-bound twin's point of view, the spaceship twin is the moving twin and so the spaceship twin will be younger than the Earth-bound twin when she eventually returns to Earth and lands (changes her axes to match the Earth-bound twin's axes) and the twins meet. However, the spaceship twin has a right to consider herself to be the stationary twin and to take the view that the Earth-bound twin boarded a planet and zoomed off. The spaceship twin says that the Earth-bound twin turned around and returned that they might meet together – the Earth bound twin did not turn around, did she? Thus, so the paradox goes, from the view of the spaceship twin, the Earth-bound twin will look and be younger.

There is a difference between the two situations of the two twins. The Earth-bound twin travels on a straight line through space-time - because her velocity never changes – she never rotates in space-time. The spaceship twin must, at the very least, change her velocity when she turns around to come back to Earth, and so she does not travel in a straight line through space-time. The bent line (through space-time) of the spaceship twin is shorter than the straight line (through space-time) of the Earth-bound twin. Since the Earth-bound twin travels along a straight line between the two events, she travels by the longest route possible through space-time between two events. Thus, she will be older than anyone who travels a shorter route (a bent route). The spaceship twin is the younger because, in space-time, a straight line is the longest distance between two points.

At the point where the spaceship twin turns around, she rotates her axes in space-time. Thus she changes the axes against which she measures the universe. At this point, all her data about the universe must be recalibrated to fit her new co-ordinate system. This recalibration includes a considerable adjustment to the age of the Earth-bound twin as seen by the spaceship twin.

Let us travel with the spaceship twin. We all know that the Earth-bound twin will age more than we do, and we keep an earthward pointing telescope trained upon her as we travel towards the distant star. As we look towards Earth from the spaceship, we see the Earth-bound twin age, but she ages slower than we age because, in our view, she is moving and therefore her clock runs slower than does ours. When we see this, we have taken proper account of the time needed for light to reach us from Earth. The Earth-bound twin's apparent slower ageing is not an effect of our being distant from her and the light from Earth taking a year or two to reach us. Now, we know that when we return and meet her, she will be older than we are, and we know that, for the year or so before we meet, she will age slower than we age as we travel back to Earth. Thus, she must appear to us to age very quickly during the time when we are turning around. If it takes us one second to reverse our spaceship and head back to the Earth, then, in that one second, we will see the Earth-bound twin age by, say, twenty years! This is not an as yet undiscovered time un-dilation effect due to the acceleration experienced by the spaceship twin. This apparent drastic ageing is due to no more than the re-aligning of the space-time axes of the spaceship twin.

The re-aligning of axes is a change of velocity and so an acceleration is involved. This is why you might hear "disappointing explanations" of the twin paradox in the form "one twin underwent acceleration". Such a "disappointing explanation" is true but disappointing. Re-aligning of axes is the explanation.

Suppose, the Earth-bound twin sees her sister approaching the Earth and decides to fly to meet her. Thus, the Earth-bound twin boards a spaceship and accelerates to match her sister's velocity. Who then is younger when they meet? Both twins have travelled "bent" lines through space-time. The one who has travelled the least bent (closest to straight) will be the oldest because her route was the longest.

All objects travel through space-time at the same speed. A stationary twin travels through space-time in the time direction whilst a moving twin travels through space-time in a "some space and some time" direction. Thus we expect that the spaceship twin will travel through less time because she is traveling through space as well as through time. However, the spaceship twin can consider herself to be stationary in space. It is only after one twin has changed her space-time axes that it can be said that she was traveling through space as well as time. To change her axes, she must change her velocity and thus "bend" the line of her travel through space-time. Only after changing her velocity does her travel line through space-time become bent. Only after changing her velocity does her travel line through space-time become shorter than the straight line of her sister. The different rates of aging are due solely to a straight line being the longest distance between two points in space-time, but which is the straight line and which is the bent line depends upon who changed their velocity.

The slowing of ageing of travellers has been experimentally observed on the Apollo missions to the moon wherein clocks on the Apollo spaceships lost 3 seconds compared with clocks on the Earth.

15.5   THE POLE AND BARN PARADOX

There is another famous seeming paradox that is associated with the special theory of relativity. It is known as the pole and barn paradox. A farmer wishes to store a 20 meter long pole in a barn, but the barn is only 10 meters long. A biology student advises the farmer that if the pole is accelerated to a velocity of $0.87c$, the length of the pole will contract to 10 meters. So, the farmer decides to so accelerate the pole to $0.87c$, and, as the shortened pole passes by the barn, he will give the pole a sideways kick into the barn. From the point of view of the farmer, the pole will fit into the barn. The paradox is that a rider on the pole sees the barn to be only 5 meters long whereas the pole is still 20 meters long, and so the pole will not fit into the barn.

What is true in one reference frame is also true in any other reference frame. Either the pole fits into the barn in both reference frames or is does not fit into the barn in both reference frames.

Let us re-word the pole and barn paradox. We have a barn that is 10 meters long and a pole that is 10 meters long. When both are stationary, the pole is the same length as the barn. When the barn is moving, the barn is seen by the rider on the pole to be shorter than the pole. When the pole is moving, the pole is seen by the stationary farmer to be shorter than the barn. This is the same as the case of the two twins moving relatively to each other who both see the other to be younger than are they. It is, as in the case of the twins, just a matter of the orientation of the space-time axes. To make an objective observation, the axes of the observers need to be aligned by a change of velocity of one of them.

EXERCISES

1. Suppose the space-time of the universe was like the surface of a higher dimensional sphere (Riemannian) in that one could travel infinitely far within it yet it is of finite extent. (There are cosmologies that accept this view.) Now suppose that two moving observers pass each other, without any acceleration, as they travel "around" this sphere; and suppose they meet again billions of years later during which time neither of them has changed velocity. They have both travelled on straight lines through space-time. Who will be the younger? Does this mean that the "spherical" view of cosmology must be incorrect?

2. A physics student borrows $100.00 from a kindly bank manager who charges the student interest at only 100% per year (as measured by the bank manager's watch). The grateful physics student arranges for the bank manager to go on a space flight at $v = 0.9c$ for a duration of one year (as measured by the physics student's watch). How much interest does the physics student pay to the bank manager when he returns from his space flight?

16

4–VECTORS

This chapter and the next are a conventional presentation of special relativity using the algebra of 4-vectors. 4-vectors were invented by Minkowski in 1907. The chapter following these two chapters is a less usual presentation of special relativity using the hyperbolic complex numbers (the algebra of space-time).

The 4-vector algebra is a central part of the conventional presentation of special relativity. It is generally agreed that, for the most part, 4-vector algebra gives the correct answers. However, there is a problem concerning the acceleration 4-vector in that it has a norm, $A \bullet A$, that is a negative number and the length of the vector is the square root of this negative norm, which is imaginary; it ought not to be this way. There is also a problem that arises when the dot product of the acceleration 4-vector and the velocity 4-vector is calculated to be zero – although dot products can be zero in space-time, it is nonsensical for them to be so. The conventional response is either to not consider the acceleration 4-vector at all (by far the most common response) or to fudge it and ignore the negativity of the norm and misuse of the zero dot product. Your author's view is that 4-vector algebra does not work properly because 4-vectors are not a *bona fide* division algebra. We therefore present the same content differently in chapter 18 of this book. We get a much better looking, but different, answer from the conventional 4-vector presentation. Fortunately, we see in Chapter 18 that it is easy to understand the difference and to come to the same answer as is produced by 4-vector calculation, but without having to fudge things.

16.1 THE STANDARD PRESENTATION OF 4-VECTORS

4-vectors do not form a proper mathematical structure in that they are not a division algebra – as with vectors in general, there is no multiplicative closure or multiplicative inverse – but they are widely used in physics because (except for the norm of the 4-acceleration and the dot product of the 4-acceleration with the 4-velocity) they work properly[1]. 4-vectors are vectors in 4-dimensional space-time. They can be based in any inertial (inertial means non-accelerating) co-ordinate system. They are different from vectors in the 4-dimensional Euclidean space, $\mathbb{R}^4$, in that they have a different inner product and norm and they take account of the different units that humankind uses to measure space and time. However, they are similar to vectors in the 4-dimensional Euclidean space in that they have four components that are just four real numbers and thus 4-vectors are in $\mathbb{R}^4$ rather than in a geometric space derived from a finite group. 4-vectors are just vectors in 4-dimensional space-time (any non-accelerating velocity less than c). It is important to realize that a "vector algebra" depends on the type of empty space within which the vectors lie; the dot product is different in different spaces. Differentiation of 4-vectors is done, as with the familiar vectors in 3-dimensional Euclidean space, by differentiating each component.

Not all Lorentz invariant physical laws are expressible with 4-vectors; some use scalars, some require 4-tensors, and particle physicists also use spinors. However, if a physical law can be written in 4-vector notation, then that physical law is guaranteed to be Lorentz invariant (invariant under rotation in space-time).

We will write 4-vectors either horizontally or vertically to suit appearance. In this book, there is no difference of meaning in using horizontal or vertical notation. (In some other books, there is a difference.) 4-vectors are written as: $\overline{\overline{A}} = [a_x \quad a_y \quad a_z \quad a_t]$, but, in order to take account of the units, we include the speed of light, c, in the final co-ordinate[2]. Note that the leftmost three components are

---

[1.] Just because something works properly, they use it – typical of physicists.
[2.] There is no standard notation denoting 4-vectors. The reader might come across various forms such as: A_4, $\mathbb{A}$.

the spatial components associated with 3-dimensional space and the rightmost component corresponds to the time component[3]. Thus, the displacement vector is written as: $R = [x \quad y \quad z \quad ct]$. Note that the physical dimensions (the units in which they are measured) of each component in the 4-vector are now the same; they are all of mass-dimension length (that is they are measured in meters). We write this as:

$$[x] = L^1, \quad [y] = L^1, \quad [z] = L^1, \quad [ct] = \left[\frac{L}{T} T\right] = L^1 \qquad (16.1)$$

We see that multiplying time by a velocity (of light) produces a length. As with "normal" vectors, a 4-vector can be thought of as an arrow, but, in the case of 4-vectors, it is an arrow in space-time (that is the space-time between the 45° asymptotes) rather than in space.

The length of a 4-vector is calculated as: *Length* $=$ $\sqrt{c^2 t^2 - x^2 - y^2 - z^2}$ [4]. The norm of a 4-vector is just its length squared. The length is also referred to as the invariant interval – it is an interval (distance) in space-time that is invariant under rotation in space-time – it is the length of the displacement vector, and so it will be invariant under rotation. We denote the invariant interval by the Greek letter tau, τ:

$$\tau^2 = c^2 t^2 - x^2 - y^2 - z^2 = \tau'^2 = c^2 t'^2 - x'^2 - y'^2 - z'^2 \qquad (16.2)$$

The dot product (inner product) of 4-vectors is:

$$\overline{\overline{A}} \bullet \overline{\overline{B}} = \begin{bmatrix} a_x \\ a_y \\ a_z \\ ca_t \end{bmatrix} \bullet \begin{bmatrix} b_x \\ b_y \\ b_z \\ cb_t \end{bmatrix} = c^2 a_t b_t - a_x b_x - a_y b_y - a_z b_z \qquad (16.3)$$

Note that:

$$\overline{\overline{A}} \bullet \overline{\overline{B}} = \overline{\overline{B}} \bullet \overline{\overline{A}} \qquad (16.4)$$

---

[3]. In some notations, it is the leftmost component that corresponds to the time component.

[4]. The reader might find that in some textbooks the length is given as $dist = \sqrt{x^2 + y^2 + z^2 + (ict)^2}$ where the imaginary square root of minus-one is used to avoid the minus signs in the distance function.

One of the disadvantages of using 4-vectors is that the notation does not clearly indicate that the component values are restricted by the requirement that:

$$c^2 t^2 > x^2 + y^2 + z^2 \tag{16.5}$$

When we use the space-time trigonometric functions {cosh( ), sinh( )} in a rotation matrix, this restriction is automatic. This notational failing has led to people postulating some kind of space-time in which the spatial distance is greater than the time distance and thus the space-time distance is a Euclidean imaginary number given by:

$$\sqrt{c^2 t^2 - x^2 - y^2 - z^2} = \sqrt{-d^2} = id \tag{16.6}$$

The number $\hat{i}$ is from the Euclidean complex numbers algebra, $\mathbb{C}$, of the 2-dimensional Euclidean plane. It does not exist in the hyperbolic complex numbers algebra, $\mathbb{S}$, of space-time. There is no such thing as space, or space-time, with imaginary distance. Because of these restrictions, the arrow that is the 4-vector can point in only directions that are within the 45° asymptotes (limiting velocity) of space-time (think velocity 4-vector). This means that there is a problem with the idea that the 4-acceleration 4-vector is orthogonal to the velocity 4-vector. The problem, as we will see in chapter 18, derives from the misapplication of the dot product to the idea of orthogonality.

4-vectors have norm (length squared) and inner-product (dot product) that is unchanged (invariant) under rotation in space-time, but the components of the vector(s) change - components of vectors always change under rotation and lengths of vectors are always invariant under rotation. The notation is a little confusing because we are using four components and we normally set our velocity to be along one of the spatial axes. We have the "counterclockwise" rotation:

$$\begin{bmatrix} 1 & 0 & 0 & 0 \\ 0 & 1 & 0 & 0 \\ 0 & 0 & \gamma & -\gamma\dfrac{v}{c} \\ 0 & 0 & -\gamma\dfrac{v}{c} & \gamma \end{bmatrix} \begin{bmatrix} x \\ y \\ z \\ ct \end{bmatrix} = \begin{bmatrix} x \\ y \\ \gamma z - \gamma\dfrac{v}{c}ct \\ -\gamma\dfrac{v}{c}z + \gamma ct \end{bmatrix} \tag{16.7}$$

Where we see that we are carrying two space dimensions along for the ride. Special relativity is essentially a theory of 2-dimensional space-time – the hyperbolic trigonometric functions are 2-dimensional. Physicists feel the need to adapt special relativity to the 4-dimensional space-time we see, and so they throw in two inert space dimensions that have nothing to do with the math and come along for no more than the ride. The 4-dimensional space-time we see is dealt with by the Lorentz group which we consider in a later chapter.

16.2   DIFFERENTIATION OF 4-VECTORS

Within 4-vector algebra, it is normal to differentiate, not with respect to time, but with respect to the invariant interval, $\tau = \sqrt{c^2 t^2 - x^2 - y^2 - z^2}$. There is no great mathematical reason for this; we could differentiate with respect to time if we wanted to (or space), but the result would be complicated by the fact that time is not an invariant under Lorentz transformation (rotation in space-time) – there is time dilation. We differentiate to compare how one variable changes with respect to another variable, and it simplifies things if the variable against which we do the comparing is not changing (dilating). Since time changes with space-time rotation in a predictable way (time dilation), we could do the calculations necessary to adjust our differentiation. We don't do this because it is easier, and felt to be aesthetically more elegant, to differentiate with respect to τ, which is, of course, the invariant length of the 4-vector. So, we are differentiating each component of the 4-vector with respect to the length of the 4-vector to see how these components change under a Lorentz transformation (space-time rotation) compared to the unchanging length of the vector. Of course, for a stationary observer, the length of her 4-vector is just the same as the time component and so, for the stationary observer $\tau = t$. This is why in Newtonian mechanics we differentiate with respect to time when calculating velocity and acceleration.

NOTE
When we differentiate in the complex plane, $\mathbb{C}$, we have to take account of the Cauchy Riemann equations. Your author is of the opinion that we have to do the same in the space-time, $\mathbb{S}$, of the hyperbolic complex numbers and so we will not be differentiating with respect to τ in the Chapter 18 of this book.

The reader will recall from a previous chapter that (9.33) & (9.35):

$$\frac{dt}{d\tau} = \frac{1}{\sqrt{1 - \dfrac{v^2}{c^2}}} = \gamma = \cosh \chi$$

$$\frac{dz}{d\tau} = \frac{v}{\sqrt{1 - \dfrac{v^2}{c^2}}} = v\gamma = \sinh \chi$$

(16.8)

In practice, we calculate by differentiating with respect to τ and then multiplying the result by $\dfrac{dt}{dt} = 1$. This gives us $\dfrac{dt}{dt}\dfrac{d(\ )}{d\tau} = \dfrac{dt}{d\tau}\dfrac{d(\ )}{dt} = \gamma\dfrac{d(\ )}{dt}$. The effect of this is to "drop" the answer on to the time axis of the stationary observer (think trigonometric functions are projections on to an axis and $\gamma = \cosh\chi$). Thus we calculate how the object of our calculation appears to a stationary observer – which is what we usually want.

16.3 4-VELOCITY

The position 4-vector (displacement 4-vector) in space-time is: $\overline{\overline{R}} = [x \quad y \quad z \quad ct]$. We differentiate this vector with respect to τ to get the velocity 4-vector (also known as the 4-velocity):

$$\overline{\overline{U}} = \left[\frac{dx}{d\tau} \quad \frac{dy}{d\tau} \quad \frac{dz}{d\tau} \quad \frac{d(ct)}{d\tau}\right] = \frac{d\overline{\overline{R}}}{d\tau}$$

(16.9)

Multiply by $\dfrac{dt}{dt}$ and extract γ:

$$\overline{\overline{U}} = \left[\frac{dt}{d\tau}\frac{dx}{dt} \quad \frac{dt}{d\tau}\frac{dy}{dt} \quad \frac{dt}{d\tau}\frac{dz}{dt} \quad \frac{dt}{d\tau}\frac{d(ct)}{dt} \right]$$

$$= \gamma \left[\frac{dx}{dt} \quad \frac{dy}{dt} \quad \frac{dz}{dt} \quad c \right] \qquad (16.10)$$

$$= \gamma \left[\vec{v} \quad c \right]$$

Where we have introduced the 3-dimensional spatial velocity vector, $\vec{V}$.

The 4-velocity is the familiar velocity in three dimensional space fitted into a 4-vector together with the velocity through time, which has value c = *speed of light*. The concept of velocity through time is a difficult one. The reader should bear in mind that time, like space, can be travelled through.

The magnitude of the 4-velocity is the length of the 4-velocity 4-vector and is given by the square root of its norm:

$$\sqrt{\overline{\overline{U}} \bullet \overline{\overline{U}}} = \sqrt{\gamma^2 \left[\vec{v} \quad c \right] \bullet \left[\vec{v} \quad c \right]}$$

$$= \sqrt{\frac{1}{1 - \dfrac{\vec{v} \bullet \vec{v}}{c^2}} (c^2 - \vec{v} \bullet \vec{v})} \qquad (16.11)$$

$$= c$$

Thus, regardless of the magnitude of the 3-dimensional spatial velocity, $\vec{V}$, through space, the 4-velocity is always c, including for a stationary observer. Yes, the stationary observer is traveling through time at the speed of light. The 4-velocity is the velocity through space-time. Readers who wish to can think of the 4-velocity as the tangent vector that is tangent to our path through space-time.

So, stationary observers travel through space-time at the speed of light – now there's a thing of import! Since the length of a 4-vector is invariant under rotation in space-time, all observers travel through space-time at the speed of light. The only difference is direction. Some observers (stationary ones) travel in the time only direction. Other observers travel in a direction that is part space and part time – this concept rather blows the mind when one first meets it. And so, the speed of light is the speed, through space-time, of everything and not the speed of only light. That is worth emphasizing:

Everything travels through space-time at the speed of light.

16.4 4-ACCELERATION

We continue to follow the conventional presentation. We have above:

$$\overline{\overline{U}} \bullet \overline{\overline{U}} = c^2 \qquad (16.12)$$

Since c is just a number, if we differentiate with respect to τ we will get zero:

$$2 \frac{\overline{\overline{}}}{} \bullet \overline{\overline{U}} = 2\overline{\overline{A}} \bullet \overline{\overline{U}} = 0 \qquad (16.13)$$

This conventional presentation is confusing because, although the 4-vector dot product in space-time can be zero (see the unrestricted way it is defined), it cannot be equal to the cosh( ) function because the cosh( ) function is never zero. (It's the failure to acknowledge the restrictions that causes the problem.) Thus the concept of orthogonality as a measure of the angle between two 4-vectors is inapplicable. It is, however, part of the standard presentation of 4-vectors, and so we will proceed. The convention is that the 4-acceleration is orthogonal (in space-time) to the 4-velocity. (What really happens is that, for a stationary observer, the acceleration vector points in the space-direction while velocity vector points in the time direction.) We will emphasise that:

4-acceleration is orthogonal (perpendicular) to 4-velocity

Thus, accepting the convention that orthogonality is equivalent to perpendicularity, we see that it is not possible to increase the 4-velocity (increase the length of the velocity vector) but only to redirect it (pull it to one side). Of course, this is what we are accustomed to; when we accelerate (change velocity), we rotate the velocity vector in space-time – pull it sideways rather than lengthen it. Now, there's a profound insight!

Although the deduction of the insight is nonsense, the insight is correct for a stationary observer, and we will see in Chapter 18 that "4-acceleration" is "perpendicular" to "4-velocity" in the space-time sense for a stationary observer.

Aside: The 4-vector dot product is zero when the magnitude of the spatial part of the 4-vector is of equal magnitude to the time part. This is when $\cosh \chi = \sinh \chi$. This is never the case since $(\cosh^2 \chi - \sinh^2 \chi = 1)$.

In a previous chapter, we pondered over why humankind were stuck with a velocity through space-time ($h = 1$) that was constant. We pointed out that this constant is the speed of light, at which all objects travel through space-time. We now understand this a little better. It is a property of space-time that 4-acceleration (the space-time type of acceleration) is perpendicular to 4-velocity. Thus, the magnitude of the 4-velocity cannot be changed; 4-velocity must be constant. This is why the speed of light is a physical constant. The speed of light is not the speed of only light. It is the speed of everything in space-time, and it is constant because the 4-acceleration is perpendicular to the 4-velocity (at least for the stationary observer – we are all stationary observers relative to ourselves). Let me emphasize that:

The speed of light is constant because the 4-acceleration is perpendicular to 4-velocity in space-time.

That's one way to get a physical constant!

We differentiate the velocity 4-vector to get the acceleration 4-vector called 4-acceleration:

$$\overline{\overline{A}} = \left[\frac{d^2x}{d\tau^2} \quad \frac{d^2y}{d\tau^2} \quad \frac{d^2z}{d\tau^2} \quad \frac{d^2ct}{d\tau^2} \right] = \frac{d\overline{\overline{U}}}{d\tau} = \gamma \frac{d\overline{\overline{U}}}{dt}$$

$$= \gamma \frac{d}{dt} \left(\gamma \left[\frac{dx}{dt} \quad \frac{dy}{dt} \quad \frac{dz}{dt} \quad c \right] \right)$$

$$= \gamma \left(\frac{d\gamma}{dt} \left(\left[\frac{dx}{dt} \quad \frac{dy}{dt} \quad \frac{dz}{dt} \quad c \right] \right) + \gamma \frac{d}{dt} \left(\left[\frac{dx}{dt} \quad \frac{dy}{dt} \quad \frac{dz}{dt} \quad c \right] \right) \right)$$

$$= \gamma \left[\frac{d\gamma}{dt} \vec{v} + \gamma \vec{a'} \quad \frac{d\gamma}{dt} c \right]$$

(16.14)

Where we have introduced the 3-dimensional spatial acceleration vector, $\vec{a'}$. Note that this $\vec{a'}$ is the acceleration of a moving body as seen by a stationary observer. It is of lesser magnitude than the acceleration, $\vec{a_0}$, that would be felt by an observer moving with the accelerating body (which is what a stationary observer would see if the accelerating body were stationary).

We have:

$$\frac{d\gamma}{dt} = \frac{d}{dt}\left(1 - \frac{\vec{v} \bullet \vec{v}}{c^2}\right)^{-\frac{1}{2}}$$

$$= -\frac{1}{2}\left(1 - \frac{\vec{v} \bullet \vec{v}}{c^2}\right)^{-\frac{3}{2}}\left(-\frac{\left(2\vec{v} \bullet \frac{d\vec{v}}{dt}\right)}{c^2}\right) \qquad (16.15)$$

$$= \gamma^3 \frac{(\vec{v} \bullet \vec{a'})}{c^2}$$

Of course, for a stationary observer, $\gamma = 1$, $\dfrac{d\gamma}{dt} = 0$, and the 4-acceleration is just:

$$\overline{\overline{A}} = \begin{bmatrix} \vec{a'} & 0 \end{bmatrix} = \begin{bmatrix} \vec{a_0} & 0 \end{bmatrix} \qquad (16.16)$$

as we think it should be. We see that the 4-acceleration of special relativity reduces to the purely spatial 3-dimensional acceleration vector of Newtonian mechanics when the velocity of the observer is zero.

There is a problem with $\overline{\overline{A}} = \begin{bmatrix} \vec{a'} & 0 \end{bmatrix}$. That problem is that:

$$\overline{\overline{A}} \bullet \overline{\overline{A}} = \begin{bmatrix} \vec{a'} & 0 \end{bmatrix} \bullet \begin{bmatrix} \vec{a'} & 0 \end{bmatrix} = 0^2 - \vec{a'}^2 \leq 0 \qquad (16.17)$$

This cannot be correct because the value of the acceleration is the square root of the norm and this is a negative number and so the acceleration is imaginary. (Again, it's that failure to acknowledge the restrictions that causes the problem.) This is the origin of the idea that the observer's acceleration is the negative of the norm of the 4-acceleration. We address this problem in Chapter 18, using matrices instead of 4-vectors.

There is an "interpretation" of the imaginary nature of the 4-acceleration. Thinking of space-time as that part of a flat piece of paper that is trapped between the 45° asymptotes that are the limiting velocity and noting that the velocity vector of a stationary observer points along the horizontal time axis, the acceleration vector must be vertical. Such a vector, translated back to the origin, would point straight up (or down) the space axis, which is outside of the 45° asymptotes, and which points in the imaginary direction in the Euclidean complex plane. I think this is an unimpressive "interpretation".

None-the-less, substitution gives:

$$\overline{\overline{A}} = \left[\gamma^4 \frac{(\vec{v} \bullet \vec{a})}{c^2} \vec{v} + \gamma^2 \vec{a} \quad \gamma^4 \frac{(\vec{v} \bullet \vec{a})}{c^2} c \right] \tag{16.18}$$

We now calculate the norm of the 4-acceleration. We have:

$$\overline{\overline{A}} \bullet \overline{\overline{A}} = \gamma \left[\frac{d\gamma}{dt} \vec{v} + \gamma \vec{a}' \quad \frac{d\gamma}{dt} c \right] \bullet \gamma \left[\frac{d\gamma}{dt} \vec{v} + \gamma \vec{a}' \quad \frac{d\gamma}{dt} c \right]$$

$$= \gamma^2 \left(\gamma^6 \frac{1}{c^2} (\vec{v} \bullet \vec{a}')^2 \left(1 - \frac{(\vec{v} \bullet \vec{v})}{c^2} \right) \right.$$

$$\left. - 2\gamma^4 \frac{1}{c^2} (\vec{v} \bullet \vec{a}')^2 - \gamma^2 (\vec{a}' \bullet \vec{a}') \right) \tag{16.19}$$

$$= \gamma^4 \left(-\gamma^2 \frac{1}{c^2} (\vec{v} \bullet \vec{a}')^2 - (\vec{a} \bullet \vec{a}') \right)$$

$$= \gamma^6 \left(-\frac{1}{c^2} (\vec{v} \bullet \vec{a}')^2 - \vec{a}' \bullet \vec{a}' + \frac{1}{c^2} (\vec{v} \bullet \vec{v})(\vec{a}' \bullet \vec{a}') \right)$$

Using (from 3-dimensional vector algebra) the identity:

$$(\vec{v} \bullet \vec{a}')^2 = (\vec{v} \bullet \vec{v})(\vec{a}' \bullet \vec{a}') - (\vec{v} \times \vec{a}') \bullet (\vec{v} \times \vec{a}') \tag{16.20}$$

(The cross-product of two 3-dimensional vectors was dealt with in the chapter on vectors.) We have:

$$\overline{\overline{A}} \bullet \overline{\overline{A}} = \gamma^6 \left(-\frac{1}{c^2} (\vec{v} \bullet \vec{a}')^2 - \vec{a}' \bullet \vec{a}' + \frac{1}{c^2} (\vec{v} \bullet \vec{v})(\vec{a}' \bullet \vec{a}') \right)$$

$$= \gamma^6 \left(\frac{1}{c^2} (\vec{v} \times \vec{a}) \bullet (\vec{v} \times \vec{a}') - \vec{a}' \bullet \vec{a}' \right) \tag{16.21}$$

When the 3-dimensional acceleration, $\vec{a}'$, is parallel to the 3-dimensional velocity, $\vec{v}$, we have $\vec{v} \times \vec{a}' = 0$ and:

$$\overline{\overline{A}} \bullet \overline{\overline{A}} = \gamma^6 (-\vec{a}' \bullet \vec{a}')$$

$$|\overline{\overline{A}}| = \gamma^3 \sqrt{-\vec{a}' \bullet \vec{a}'} \tag{16.22}$$

This cannot be correct (the minus sign). The square root of $\overline{\overline{A}} \bullet \overline{\overline{A}}$ is the scalar amount of 4-acceleration through space-time. We get around this incorrectitude by fudging it. We declare that the kind of acceleration we are used to (the 3-dimensional $\vec{a}$) is the negative of $\overline{\overline{A}} \bullet \overline{\overline{A}}$ and that $a_0 = -\overline{\overline{A}} \bullet \overline{\overline{A}}$. The justification for this is the minus signs in the definition of the 4-vector dot product. If we could set time to zero, this would be the result. This leads to:

$$-\overline{\overline{A}} \bullet \overline{\overline{A}} = \gamma^6 \left(\vec{a'} \bullet \vec{a'} \right)$$
$$-\left| \overline{\overline{A}} \right| = \gamma^3 \sqrt{\vec{a'} \bullet \vec{a'}} \qquad (16.23)$$
$$= \gamma^3 a'$$

This is the acceleration transformation we derived earlier:

$$a' = \frac{a_0}{\gamma^3} \qquad (16.24)$$

Where a_0 is the acceleration as measured by the observer moving with a rocket and a' is the acceleration of the moving rocket as measured by the stationary observer as it passes him.

It is customary to call $\vec{a}_0$ the proper acceleration. It is the acceleration experienced by a stationary observer as measured on an accelerating but instantaneously stationary rocket - in the instantaneous rest frame. Taking the negative of the length of the 4-acceleration gives:

$$\vec{a}_0 = \gamma^3 \sqrt{\vec{a'} \bullet \vec{a'} - \frac{1}{c^2} \left(\vec{v} \times \vec{a'} \right) \bullet \left(\vec{v} \times \vec{a'} \right)} \qquad (16.25)$$

This is the 4-vector acceleration transform. If a rocket produces $\vec{a}_0$ acceleration when it is stationary, it will appear to the stationary observer to produce $\vec{a'}$ acceleration when it is moving at $\vec{v}$. When the acceleration is parallel to the velocity, this is:

$$a' = \frac{a_0}{\gamma^3} \qquad (16.26)$$

When the 3-dimensional spatial acceleration is perpendicular to the 3-dimensional spatial velocity (think orbiting particles moving in storage rings at CERN), this is:

$$a_0 = \gamma^3 \frac{1}{c}\sqrt{a'^2c^2 - a'^2v^2}$$

$$= \gamma^3 a'\sqrt{1 - \frac{v^2}{c^2}} \qquad (16.27)$$

$$a' = \frac{a_0}{\gamma^2}$$

EXERCISES

1. Is the perpendicularity of 4-acceleration and 4-velocity a deep insight into the nature of the universe and of the origin of physical constants?

2. What is the length of the 4-vectors $[6 \quad 4 \quad 1 \quad 3]$ & $[6 \quad 4 \quad 4 \quad 3]$

3. What is the space-time angle (think dot-product and cosh( )) between the two 4-vectors $[6 \quad 4 \quad 1 \quad 3]$ & $[6 \quad 2 \quad 4 \quad 3]$?

4. What are the mass-dimensions of gamma?

5. A rocket is moving with 4-velocity $[6 \quad 2 \quad 2 \quad 4]$ as it recedes directly away from a stationary observer. It accelerates directly away from the observer in such a way that its instruments record its 4-acceleration to be $[3 \quad 1 \quad 1 \quad 2]$. What will its spatial acceleration be as measured by the stationary observer?

6. What is the norm of the 4-acceleration $[4 \quad 3 \quad 1 \quad)$?

7. What is the 4-acceleration of a spaceship with spatial acceleration $[1 \quad 2 \quad 3]$ moving at space-velocity $[2 \quad 0 \quad 1]$?

8. What is the 4-acceleration of a spaceship with 4-velocity $[12t^3 \quad 2x \quad y \quad 3z^2t)$?

17

4-MOMENTUM AND RELATIVISTIC MASS

We have seen above that the 3-dimensional spatial acceleration, $\vec{a}$, is not invariant under change of velocity. This is contradistinct from the Newtonian world view in which the acceleration is invariant. We might anticipate that the relativistic view of momentum and force will differ from the Newtonian view.

In Praise of Newton: Newtonian mechanics has successfully predicted the orbits of planets (excepting a minor discrepancy in the orbit of Mercury – which special relativity does not correct) and successfully, and delicately, landed space-probes on distant worlds. Newtonian mechanics is used by space-agencies world-wide for calculating the trajectories and orbits of their rockets and probes. Newtonian mechanics is very successful. It is so successful that the fact deserves emboldening.

The only thing wrong with Newtonian mechanics is that it is not right[1].

Aside: The first writing of the mechanics of special relativity was done by Max Planck (1858–1947) in 1906. A second writing of it was done by Lewis and Tolman in 1909[2].

[1.] CERN finds huge discrepancies between reality and Newtonian mechanics because they deal with particles moving at very close to the speed of light.

[2.] Lewis & Tolman Phil Mag 1909 Vol. 18 pp. 510–533.

Modern physics accepts that every particle in the universe has a real number (a scalar) associated with it that we call the particle's rest-mass. We denote this rest-mass as m_0. The rest-mass might be zero – it might be that such particles are outside of space-time. Special relativity does not try to explain what mass is or why everything has mass but starts by accepting as given this assumption of rest-mass. Nor does special relativity take any account of gravitation (general relativity does that). To special relativity, mass is associated with inertia.

Aside: According to quantum theory, particles have rest-masses that are proportional to Planck's constant, $\hbar$. Planck's constant plays no role in special relativity.

17.1 ABOUT MASS

We do not understand mass properly; even so, we will discuss it. We remind the reader that the mass-dimensions of space and time are the same in special relativity:

$$[L] = [T] \tag{17.1}$$

and that this means that the units used to measure energy and to measure mass are the same as the units used to measure momentum. Which is another way of saying that energy, mass, and momentum are the same kind of "stuff".

The phrase "rest-mass" is a little misleading. Everything travels through space-time (as distinct from space) at the speed of light, and so nothing is at rest. The rest-mass of a stationary particle is really the "momentum" of the particle in the time direction, which is the energy of the particle.

> *The rest-mass of a stationary particle is just its momentum in the time direction.*

The time direction of space-time is associated with the cosh( ) function. It seems that rest-mass is associated with the fact that $\cosh(0) = 1$.

From the point of view of the stationary observer, light does not travel through time (think time dilation). If a photon of light does not travel through time, it has no time component of momentum and so it has zero rest-mass. This does not mean that it has zero momentum; it will have a spatial component of momentum (which is also an energy), and so a photon of light has momentum even though it has zero rest-mass. That seems to make sense, and it seems to fit what we observe. However, there is a problem within theoretical physics associated with particles called neutrinos. Neutrinos seem to travel through space at the speed of light; they also seem to have non-zero rest-mass. There is clearly something to do with the speed of light and the rest-mass of particles that we do not understand.

Aside: Neutrinos have intrinsic spin which, when compared to the neutrino's direction of motion, allows us to associate a definite helicity with the neutrino. It seems that only left-helicity neutrinos (and right-helicity anti-neutrinos) exist. This means that as a neutrino passes by us to our right, we always see it with left-helicity. If we were to overtake the neutrino, it would appear to pass us to our left as we passed it and thus it would appear to have right-helicity, but there is no such thing as a right-helicity neutrino. Thus, it must be that we cannot overtake neutrinos, and so they must travel at the speed of light; in which case they must be massless (or so the standard story goes). Observations lead us to believe that neutrinos have mass – that's the problem within theoretical physics.

Having assumed the concept of rest-mass, we are able to define the momentum 4-vector (known as 4-momentum or as momenergy) as the product of this rest-mass and the 4-velocity:

$$\overline{\overline{P}} = m_0 \overline{\overline{U}} \tag{17.2}$$

17.2 MOMENTUM AND ENERGY CONSERVATION IN SPECIAL RELATIVITY

The 4-momentum, $\overline{\overline{P}}$, is a space-time vector. As it rotates in space-time, its components will change but its length will stay the same.

Now, every stationary observer sees within their own reference frame two conservation laws, conservation of momentum and conservation of energy. When the observer considers an event in more than one reference frame, then energy and spatial momentum become unified (as do space and time) into one entity, the 4-momentum, usually called momenergy, and there is only one conservation law – conservation of momenergy.

Momentum and energy are a special case of a general phenomena. A stationary observer sees time as separate from space. Despite this, to understand the physics of a moving physical system, the stationary observer has to unify space and time into space-time. The "God's eye view" is that space-time is not two separate things even though they will always appear to be so to a stationary observer. The same is true with force and power, with electric field and magnetic field, and with any two physical concepts that are unified by special relativity. It is true with energy and spatial momentum. To the stationary observer, in her own reference frame, energy and spatial momentum will always be two separate things and they are governed by two separate conservation laws. The "God's eye view" is that only the unification of energy with spatial momentum, called momenergy, exists and that there is only one conservation law – conservation of momenergy. The stationary observer needs to adopt the "God's eye view" if she is to understand physics within a reference frame that is moving relative to her own.

In special relativity, momenergy, also known as 4-momentum, is conserved under change of velocity. Of course it is; momenergy is the length of the 4-momentum vector. The length of a vector does not change when it is rotated. We'll say that again:

4-momentum (momenergy) is conserved under change of velocity.

Imagine two particles colliding. In a particular reference frame, prior to the collision, there is a 4-momentum 4-vector associated with each particle. The total 4-momentum is the component-wise sum of the two particle 4-momenta. It is the length (magnitude) of this total 4-momentum that is conserved through the collision. After the collision, the two particles will again each have a 4-momentum. The component-wise sum of the two post-collision 4-momenta will be equal to the pre-collision total 4-momentum.

In different reference frames, the components of the total 4-momentum will differ, but the length of the total 4-momentum will be the same. Because observers in relative motion disagree on the direction in space-time of the 4-momentum vector (its angle with the time axis), they disagree on the components of the 4-momentum vector. They agree only on its length.

So, to repeat, observers in relative motion disagree on the components of the 4-momentum but they agree on the magnitude of the 4-momentum. Of course, they do; as we rotate a vector, its components change but its length stays the same. Well, we've reiterated that enough!

For some observers, the momenergy 4-vector of a canon ball will be entirely energy; this corresponds to the canon ball being stationary. For other observers, the same momenergy 4-vector will be some energy and some spatial momentum. For all observers, the energy component will be greater than the spatial momentum component. This is exactly the same thing as the limiting velocity of the universe. The energy and momentum are as the time and space axes, and a direction in the energy-momentum space-time cannot be steeper than the limiting 45° asymptotes.

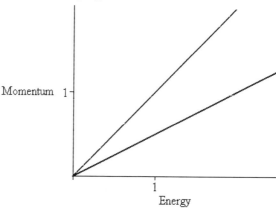

To digress from 4-vectors briefly: For a cannonball of rest-mass m_0 kilogram moving at a velocity through space-time that corresponds to the space-time angle χ, the energy of the cannonball is given by $E = m_0 \cosh \chi$ and the spatial momentum of the cannonball is given by $p = m_0 \sinh \chi$.

The momenergy 4-vector (that is the 4-momentum 4-vector) points in the same direction in space-time as the 4-velocity of the particle through space-time - this is not the same as the velocity through space.

We have:

$$\overline{\overline{P}} = m_0 \overline{\overline{U}}$$

$$= m_0 \gamma(u) \begin{bmatrix} \vec{u} & c \end{bmatrix} \tag{17.3}$$

$$= \begin{bmatrix} m_0 \gamma \vec{u} & m_0 \gamma c \end{bmatrix}$$

The spatial part of this 4-vector is:

$$\begin{bmatrix} \vec{p} & \sim \end{bmatrix} = \begin{bmatrix} \gamma m_0 \vec{u} & \sim \end{bmatrix} \tag{17.4}$$

Now, in Newtonian mechanics, we have that 3-dimensional momentum is the product of mass and velocity, $\vec{p} = m\vec{u}$, and, because in Newtonian mechanics we are concerned with only slow velocities, the mass is taken to be the rest-mass. In special relativity, we accept that the mass (not rest-mass) is not an invariant quantity under rotation in space-time (change of velocity). This contrasts with electric charge which is an invariant quantity under rotation in space-time. We therefore, looking at the momentum 4-vector above, define relativistic mass to be:

$$m' = m_0 \gamma$$

$$= m_0 \frac{1}{\sqrt{1 - \dfrac{v^2}{c^2}}} \tag{17.5}$$

Thus it is that the mass of a body seems to the stationary observer to increase with velocity.

Theoretical physicists equate the mass-dimensions of time with those of space. They also have the mass-dimensions of mass as:

$$[M] = L^{-1} = \frac{1}{L} \tag{17.6}$$

The length of a moving rod contracts, and so, looking at the mass-dimensions of mass, we ought not to be surprised that its mass increases.

Of course, the square root gives rise to both positive and negative mass, but no-one has ever observed negative mass. For that matter, no-one has ever observed negative time from the time dilation formula. It has been postulated that anti-matter might be gravitationally negative mass, but there is evidence that this is not so. Experiments by Eotvos (1848–1919) and by Dicke (1916–1997) would have detected such gravitationally negative mass, and there is astrophysical evidence that the gravitational mass of the K° meson and its anti-particle are the same to within a few parts in 10^{10}.

Aside: Within particle physics, the rest-mass of a particle is thought to derive from the Higgs mechanism. The Higgs mechanism is neither conceptually nor mathematically elegant, but no-one has invented any better way of explaining mass.

Rest-mass is approximately only 1% of the mass in the universe. The other 99% of the mass of the universe is the binding energy within the quark gluon plasma as described by quantum chromodynamics, QCD.

It is often said that the mass of a body tends to infinity as the velocity of the body through space tends to the speed of light, and so the force necessary to accelerate this body tends to infinity as the velocity through space tends to the speed of light, and so a body with non-zero rest-mass cannot accelerate through the "light-barrier". In spite of this being both often said and true, it misses the point. The point is that nothing within space can accelerate through the "light-barrier" because the speed of light is the limiting velocity. If a massless particle was moving slower than the speed of light through space (there is no such thing known), it could not be accelerated through the "light-barrier".

Special relativity does not deny the existence of particles moving faster through space than the speed of light. They would not be in space-time, but the theory does not deny their existence - it says nothing about them. Such particles are called tachyons. They have been searched for and not found[3]. Of course, "moving faster" seems to be something that can happen in only space-time.

[3.] They exist in Star Trek movies, but then so does warp speed.

Aside: One of the great unsolved mysteries of life is why the inertial mass of a body, which a body will have in the complete absence of gravity, should be exactly equal to the gravitational mass (the charge of the gravity force) of the body. Experiments by Eotvos in 1889 and again in 1922 and by Roll, Krotkov, and Dicke[4] in 1964 and by Braginski and Panov[5] in 1971 have verified this equivalence to one part in 10^{12}. It is because the gravitational mass is equal to the inertial mass that the acceleration due to gravity is equal to the strength of the gravitational field. The mystery is that the mass that is the resistance to acceleration by an electric field is the same as the mass that partakes of the gravitational field.

Now:

$$m' = m_0 \frac{1}{\sqrt{1 - \dfrac{v^2}{c^2}}}$$

(17.7)

$$= m_0 + \frac{1}{2}m_0\frac{v^2}{c^2} + \frac{3}{8}m_0\frac{v^4}{c^4}\ldots + O\left(\frac{v^6}{c^6}\right)$$

This implies:

$$mc^2 = m_0 + \frac{1}{2}m_0v^2 + \frac{3}{8}m_0\frac{v^4}{c^2}\ldots$$

(17.8)

Now the second term is what we recognise as Newtonian kinetic energy. It therefore seems, and this is verified by nuclear bombs and shining suns[6], that the relativistic energy is the whole of the series which is, of course, just:

$$E = m'c^2$$

$$= m_0\gamma c^2$$

(17.9)

Thus it is that we can take the view that rest-mass is just a "frozen" form of energy. With zero spatial-velocity, we have:

$$E = m_0c^2$$

(17.10)

---

[4] Roll, Krotkov & Dicke: Ann Phys 26, 442 (1964).
[5] Braginski & Panov: Sov Phys JEPT 34 464 (1971).
[6] The sun radiates away 4 million metric tons of mass per second.

This is the most famous of all equations. Note that the units on either side of the equation match because the mass-dimension of space is the same as the mass-dimension of time and so velocity is just a unit-less number. To put some numbers into this, a kilogram of mass is equivalent to 10^{17} joules. The Hiroshima bomb released 10^{14} joules of energy.

All forms of energy except potential energy by virtue of position in a field are equivalent to mass. However, the energy of the field has mass (which gravitates), and the gravitational field has mass (which gravitates), as does the electromagnetic field.

Aside: Pressure is a kind of energy. Thus, pressure has mass. Thus pressure gravitates. When a supernova begins to gravitationally collapse, the pressure inside the star increases. The gravitational force of contraction due to the "mass of the pressure" increases faster than the outward pushing force of the pressure itself. In very large stars, the pressure increases to the point where the inward gravitational force of the (equivalent mass) pressure itself becomes stronger than the outward pressure force. The star collapses to become a black hole. It is dragged into collapse by the strength of the pressure resisting its collapse.

From above, we have:

$$\overline{\overline{P}} = \begin{bmatrix} m_0 \gamma \vec{u} & m_0 \gamma c \end{bmatrix}$$
$$= \begin{bmatrix} \vec{p} & \dfrac{E}{c} \end{bmatrix} \tag{17.11}$$

and so we see that our momentum 4-vector is comprised of the usual 3-dimensional spatial momentum vector (three components with relativistic mass) and a scalar (real number) energy component in the place of the time component. In general, 4-vectors have this structure. They have three components that correspond to the Newtonian physics and a fourth component that sits in the place of time. This is unification of 3-dimensional space and time into 4-dimensional space-time.

Note that, with $c = 1$, the ratio of the spatial component of the momentum 4-vector to the time component (that is the ratio of the energy to the spatial momentum) is just the ratio of the spatial velocity to the speed of light.

$$\overline{\overline{P}} = m_0 \gamma \left[\vec{u} \quad c \right]$$

$$v = \frac{p}{E}$$

(17.12)

This is just $v = \tanh \chi = \dfrac{\sinh \chi}{\cosh \chi} = \dfrac{p}{E}$.

Using $\overline{\overline{U}} \cdot \overline{\overline{U}} = c^2$, we have:

$$\overline{\overline{P}} \cdot \overline{\overline{P}} = m_0^2 \overline{\overline{U}} \cdot \overline{\overline{U}} = m_0^2 c^2 = \frac{E^2}{c^2} - p^2$$

(17.13)

Leading to:

$$E^2 = c^2 p^2 + m_0^2 c^4$$

$$m_0^{\,2} = E^2 - p^2$$

(17.14)

This is just $m_0^2 \left(\cosh^2 \chi - \sinh^2 \chi \right) = m_0^2$.

With $c = 1$, the rest-mass of a particle is given by the energy and momentum components of its 4-momentum by (17.14). Consider a particle at rest relative to a stationary observer; its momenergy 4-vector points in the time direction. Its rest-mass is all energy and zero spatial momentum. Now accelerate that particle; the momenergy 4-vector is rotated in space-time and the particle's mass becomes partly energy and partly spatial momentum. As the momenergy 4-vector rotates, the particles kinetic energy increases and the particles spatial momentum increases, but, because the measure of space-time distance includes a minus sign (we have that the increase in energy minus the increase in spatial momentum is constant), the magnitude of the momenergy (the length of the 4-momentum vector) stays the same. The equation $E^2 - p^2 = \text{constant}^2$ is the equation of a hyperbola, of course. This "space-time length of the momenergy 4-vector" is the particles rest-mass; it is the constant hyperbolic-radius of the hyperbola. Should we emphasize that?

The rest-mass of a particle is the space-time length of its momentum 4-vector.

Clearly, the non-zero rest-mass derives from the nature of the cosh( ) function. So, why do different particles have different rest-masses then?

Aside: Within quantum mechanics, we have the operators: $E \rightarrow i\hbar \dfrac{\partial}{\partial t}$ & $p \rightarrow -i\hbar \dfrac{\partial}{\partial x}$. If we put these into the above relativistic equation operating on the field ψ, we get:

$$E^2 = c^2 p^2 + m_0^2 c^4$$

$$\left(i\hbar \frac{\partial}{\partial t} \right)^2 \psi = c^2 \left(-i\hbar \frac{\partial}{\partial x} \right)^2 \psi + m_0^2 c^4 \psi \quad (17.15)$$

$$\hbar^2 \frac{\partial^2 \psi}{\partial t^2} - \hbar^2 c^2 \frac{\partial^2 \psi}{\partial x^2} + m_0^2 c^4 \psi = 0$$

With $c = \hbar = 1$

$$\frac{\partial^2 \psi}{\partial t^2} - \frac{\partial^2 \psi}{\partial x^2} + m_0^2 \psi = 0 \quad (17.16)$$

A mere notational change gives:

$$\left(\partial_\mu \partial^\mu + m^2 \right) \varphi = 0 \quad (17.17)$$

This is the quite famous Klein-Gorden equation of quantum field theory.

17.3 COLLIDING STICKY BUDS AND THE CENTER OF MASS REFERENCE FRAME

It is said that during the summer of 1705, Isaac Newton observed a stationary sticky bud being hit by a moving sticky bud. The two sticky buds stuck together and moved off at a velocity that was slower than the initial velocity of the moving sticky bud. Note that evolution has arranged that sticky buds are all of the same rest-mass, m_0. Let us see how Newton analysed the collision. Let u be the initial velocity of the moving sticky bud and let v be the velocity of the sticky bud pair. Newtonian conservation of linear momentum and the Newtonian invariance of mass means that we have:

$$m_0 u = 2 m_0 v \Rightarrow v = \frac{u}{2} \quad (17.18)$$

However, Newtonian kinetic energy is not conserved. We have:

$$KE_{initial} = \frac{1}{2}m_0 u^2$$

$$KE_{final} = \frac{1}{2}2m_0 v^2 = \frac{1}{2}2m_0\left(\frac{u}{2}\right)^2 = \frac{1}{4}m_0 u^2$$

(17.19)

Newton concluded that the collision of the two sticky buds converted kinetic energy into heat and noise.

It is said that during the summer of 1905, Albert Einstein observed a stationary sticky bud being hit by a moving sticky bud. The two sticky buds stuck together and moved off at a velocity that was slower than the initial velocity of the moving sticky bud. Note that evolution had continued to arrange that sticky buds are all of the same rest-mass, m_0. Let us see how Einstein analysed the collision. Relativistic conservation of linear momentum and the relativistic variance of mass means that we have:

$$\frac{m_0 u}{\sqrt{1-\dfrac{u^2}{c^2}}} = \frac{M_0 v}{\sqrt{1-\dfrac{v^2}{c^2}}}$$

(17.20)

Relativistic conservation of energy is:

$$\frac{m_0 c^2}{\sqrt{1-\dfrac{u^2}{c^2}}} + \frac{m_0 c^2}{\sqrt{1-\dfrac{0^2}{c^2}}} = \frac{M_0 c^2}{\sqrt{1-\dfrac{v^2}{c^2}}}$$

(17.21)

We have to allow for the increase or loss of mass - heat is energy, and so it has mass – so we put M_0 as the mass of the sticky bud pair. We cannot assume $M_0 = 2m_0$. Solving these equations leads to:

$$\frac{v}{u} = \frac{1}{1+\sqrt{1-\dfrac{u^2}{c^2}}}$$

$$\frac{M_0}{m_0} = \sqrt{2}\left(1+\frac{1}{\sqrt{1-\dfrac{u^2}{c^2}}}\right)^{\frac{1}{2}}$$

(17.22)

Note that if $u = 0$, $M_0 = 2m_0$ but not if $u > 0$. That $M_0 > 2m_0$ if $u > 0$ is seen by Einstein to be the mass of the heat energy in the sticky bud pair. To recap: Newton saw a non-conservation of kinetic energy; Einstein saw a conservation of total energy; they both saw a conservation of spatial-momentum.

17.4 CENTER OF MASS REFERENCE FRAME

It is easier to do the analysis in a reference frame in which, prior to the collision, the total momentum is zero. Such a reference frame is called the center of mass reference frame. It is really the center of momentum frame, but it's only a name. Since linear momentum is conserved, then the center of momentum reference frame is the reference frame in which the sticky bud pair will be stationary after the collision and in which the (equal mass) two sticky buds prior to the collision will be moving at the same speed in opposite directions. It is usual to do the analysis in the center of mass frame.

Now, using the above, we have in the center of momentum reference frame that the sticky bud pair is stationary ($\gamma = 1$), and thus its spatial momentum is zero and the time part of the 4-momentum is $E = Mc$, and so the center of mass (or momentum) reference frame has the 4-momentum as: $[Mc \quad 0 \quad 0 \quad 0]$ with norm M^2c^2. The norm is invariant under Lorentz transformations, and so we know that the norm of the 4-momentum in the original (not center of mass) reference frame will have a norm of the same value. Within the original system, the 4- momentum was: $[\gamma_u cm_0 + cm_0 \quad 0 \quad 0 \quad \gamma_u mu]$. Equating the norms gives:

$$M^2 c^2 = \left(y_u cm_0 + cm_0 \right)^2 - \left(\gamma_u mu \right)^2 \qquad (17.23)$$

Which leads, far more simply, to the equations (17.22).

17.5 4-FORCE

In Newtonian mechanics, we have two definitions of force (which never was a good idea). We define force as the product of mass and acceleration, $F = m_0 a$, and we define force as the rate of change of momentum with respect to time, $F = \dfrac{dp}{dt}$. In Newtonian mechanics, these two definitions are completely equivalent. In special relativity, we have only one definition, which is based on the rate of change of 4-momentum with respect to the invariant interval, τ. We define 4-force as the 4-vector:

$$\overline{\overline{F}} = \frac{d\overline{\overline{P}}}{d\tau} \tag{17.24}$$

This definition fits with the electromagnetic force. This definition is such that the momentum remains unchanged unless there is a force. We have:

$$\begin{aligned}
\overline{\overline{F}} &= \frac{d\overline{\overline{P}}}{d\tau} \\
&= \frac{d}{d\tau}\left(m_0 \overline{\overline{U}}\right) \\
&= m_0 \overline{\overline{A}} + \gamma \overline{\overline{U}} \frac{dm_0}{dt}
\end{aligned} \tag{17.25}$$

Now, using : $\overline{\overline{A}} = \gamma\left[\dfrac{d\gamma}{dt}\vec{v} + \gamma\vec{a}' \quad \dfrac{d\gamma}{dt}c\right]$ and $\overline{\overline{U}} = \gamma(v)\left[\vec{v} \quad c\right]$ leads to:

$$\overline{\overline{F}} = m_0\left[\gamma\frac{d\gamma}{dt}\vec{v} + \gamma^2\vec{a}' \quad \gamma\frac{d\gamma}{dt}c\right] + \gamma^2\left[\vec{v} \quad c\right]\frac{dm_0}{dt} \tag{17.26}$$

using: $\dfrac{d\gamma}{dt} = \gamma^3 v\dfrac{dv}{dt}$, we get

$$\overline{\overline{F}} = \gamma\left[m_0\gamma^3 v\frac{dv}{dt}\vec{v} + m_0\gamma\vec{a}' + \gamma v\frac{dm_0}{dt} \quad \left(m_0\gamma^3 v\frac{dv}{dt} + \gamma\frac{dm_0}{dt}\right)c\right] \tag{17.27}$$

Taking m_0 to be a constant and $\dfrac{dv}{dt} = a'$:

$$\begin{aligned}
\overline{\overline{F}} &= m_0\left[\gamma^4 v a' \vec{v} + \gamma^2\vec{a}' \quad \gamma^4 v a' c\right] \\
&= \gamma^2 m_0 a'\left[\gamma^2 v^2 + 1 \quad \gamma^2 v c\right] \\
&= \gamma^4 m_0 a'\left[1 \quad v c\right]
\end{aligned} \tag{17.28}$$

Recalling that $\gamma^3 a' = a_0$

$$\overline{\overline{F}} = \gamma^4 m_0 a'\left[1 \quad v c\right] = \gamma m_0 a_0\left[1 \quad v c\right] \tag{17.29}$$

Giving us that the 3-dimensional force is:

$$f = \gamma m_0 a_0 \qquad (17.30)$$

Thus, we see that the Newtonian $F_0 = m_0 a_0$ is modified in special relativity.

Note that the 4th component of the 4-force 4-vector, with $c = 1$, is just the velocity, v, and so the total 4-force will be dependent upon velocity just as the magnetic part of the total electromagnetic force is dependent upon velocity. This is telling us that a 3-dimensional velocity-independent force cannot be Lorentz invariant, and so cannot be a physical law. Within special relativity, almost everything depends upon velocity (time, length, acceleration, mass...), and so we ought not to be surprised that force depends upon velocity. The Lorentz force of electromagnetism (see below) is a typical relativistic force.

We also have:

$$\overline{\overline{F}} = \frac{d\overline{\overline{P}}}{d\tau} = \gamma \frac{d}{dt}\left[\vec{p} \quad \frac{E}{c}\right]$$
$$= \gamma\left[\frac{d\vec{p}}{dt} \quad \frac{1}{c}\frac{dE}{dt}\right] = \gamma\left[\vec{F} \quad \frac{1}{c}power\right] \qquad (17.31)$$

Again, we have the characteristic structure of a 4-vector. That is we have a 3-dimensional spatial vector and a scalar that corresponds to the time component. In this case, we have the 3-dimensional force vector and the power scalar. The reader should observe that unifying 3-dimensional space with time into 4-dimensional space-time automatically unifies lots of other concepts together – such as force and power. If one thinks of the units in which different things are measured and, within these units, one sets the mass-dimension of time equal to the mass-dimension of length, then one can see why unifying space and time leads to these unifications. One can equally see why, to the stationary observer, the unifications break because the stationary observer sees space and time as separate.

Immediately above, we see that the time component of 4-force is, with $c = 1$ and not worrying about the gamma, given by $\frac{dE}{dt}$. Within the unifying spirit of space-time, we ought to have the spatial component of 4-force given by $\frac{dE}{dr}$. This is exactly the concept of force that is used in particle physics. It is based on the assumption that the universe seeks to drop into the state with the lowest possible energy and so seeks to lower the energy by moving particles together (or apart). Thus, we might say, spatial force is the rate of change of energy with respect to spatial separation. Looking at the equation immediately above, we would generalize this into: space-time force is rate of change of energy with respect to space-time separation. Perhaps, we were unwise to define force as rate of change of momentum with respect to τ and ought to have defined it as rate of change of energy with respect to τ (which is spatio-temporal separation). But then, within special relativity, we have the unification of energy and momentum into momenergy, and so we effectively did that anyway. There's the answer; force is rate of change of momenergy with respect to the invariant interval, τ.

Within the universe, we observe four different types of force. These are: the gravitational force, the electromagnetic force, the weak force, and the strong force. If force is rate of change of momenergy with respect to spatio-temporal separation, we must have four types of space or four types of momenergy. The finite groups give us different types of space.

17.6 ELECTROMAGNETISM

The 3-dimensional vector for electrical current is:

$$\vec{j} = \rho\vec{u} \qquad (17.32)$$

Wherein, ρ is the current density. We will see later that the current density is not Lorentz invariant because of length contraction effects and that we have $\rho = \gamma\rho_0$. The 4-current 4-vector is:

$$\bar{J} = \rho_0\bar{\bar{U}} = \rho_0\gamma\left[\vec{u} \quad c\right] = \left[\rho\vec{u} \quad \rho c\right] = \left[\vec{j} \quad \rho c\right] \qquad (17.33)$$

The electromagnetic force (Lorentz force) 4-vector is:

$$\bar{\bar{F}} = q\gamma\left[-\left(\vec{E} + \frac{1}{c}\vec{v}\times\vec{B}\right) \quad \frac{1}{c}\vec{E}\bullet\vec{v}\right] \qquad (17.34)$$

We will deal with electromagnetism separately later.

WORKED EXAMPLES

1. What is the momenergy 4-vector of a particle whose kinetic energy is four times its rest energy?

    Ans: The total energy of the particle is $5m$. We have:

    $$p^2 = E^2 - m_0^2 \Rightarrow p^2 = \left(5m_0\right)^2 - m_0^2$$
    $$= 24m_0^2 \Rightarrow p = \sqrt{24}m_0 \qquad (17.35)$$

    Since the direction of the particle in 3-dimensional space is not specified, we are free to choose any set of the three spatial components of the 4-vector that (using Pythagoras) give the total spatial momentum as $\sqrt{24}m_0$. We choose:
    $\begin{bmatrix} 5m_0 & 0 & 0 & \sqrt{24}m_0 \end{bmatrix}$.

2. The same particle as in 1 is observed by a passing alien in a spaceship to have its kinetic energy equal to its mass. In the alien's frame of reference, what is the momenergy 4-vector?

    Ans: In the alien's frame of reference, the total energy of the particle is $2m$. This gives $p = \sqrt{3}m$ and, since no directions are specified, the 4-vector in the alien's reference frame is, by somewhat eccentric choice: $[2m \quad m \quad m \quad m]$. Note, if the directions were specified, we would have no choice.

3. In a particular reference frame, two particles collide with 4-momenta $[4 \quad 1 \quad 0 \quad 2]$ & $[3 \quad 0 \quad 1 \quad 2]$. After the collision, the particles the 4-momenta of the particles is measured and found to be $[3 \quad 1 \quad 1 \quad 1]$ & $[2 \quad 1 \quad 1 \quad 1]$. Is there some energy and momentum missing? If so, what is missing?

    Ans: The total 4-momenta prior to the collision was $[7 \quad 1 \quad 1 \quad 4]$ which has length $l = \sqrt{7^2 - 1^2 - 1^2 - 4^2} = \sqrt{31}$. This must be conserved, but the total 4-momentum after the collision is $[2 \quad 2 \quad 2 \quad 2]$ with length $\sqrt{13}$. There is something missing. Because we are in a particular reference frame, each component of the 4-momentum is conserved, and we can say that there must be some unseen particle with 4-momentum $[2 \quad -1 \quad -1 \quad 2]$. The length of this 4-momentum vector is $\sqrt{-2}$; clearly, something is amiss. What is wrong?

4. In a particular reference frame, the 4-momentum of a par-
    ticle is [5 1 0 0]. What is its 4-momentum as seen by
    an observer moving at half the speed of light relative to the
    particular reference frame?

    Ans: Change of velocity is a 2-dimensional rotation, and so
    we can ignore the two zeros in the given 4-vector. We have:

$$\begin{bmatrix} \gamma & v\gamma \\ v\gamma & \gamma \end{bmatrix}\begin{bmatrix} 5 \\ 1 \end{bmatrix} = \begin{bmatrix} 5\gamma + v\gamma \\ 5v\gamma + \gamma \end{bmatrix} = \begin{bmatrix} 5\dfrac{2}{\sqrt{3}} + \dfrac{1}{2}\dfrac{2}{\sqrt{3}} \\ 5\dfrac{1}{2}\dfrac{2}{\sqrt{3}} + \dfrac{2}{\sqrt{3}} \end{bmatrix} = \begin{bmatrix} \dfrac{11}{\sqrt{3}} \\ \dfrac{7}{\sqrt{3}} \end{bmatrix} \qquad (17.36)$$

EXERCISES

1. What is the 4-force associated with the 4-momentum
    $[5t^3 \quad x \quad y \quad z^2t]$?

2. What is the spatial part of the force associated with the 4-mo-
    mentum $[5t^3 \quad x \quad y \quad z^2t]$?

3. If, within a physical system, energy is conserved, then the
    rate of change of energy is zero, and so force is zero unless
    energy is not conserved. Thus, a force must be associated
    with a change of energy. We see this with the change of ki-
    netic energy resulting from an accelerating force. Is this why
    we need a force to increase velocity? And what is this to do
    with mass?

4. There is a significant difference between rotation in Euclid-
    ean space and rotation in space-time. In space, I need a force
    to start a body rotating, but once it is rotating, it continues
    rotating without any force. In space-time, as soon as I stop
    applying a force to a body, it stops rotating (accelerating) and
    stays pointing in the direction (velocity) in which it pointed
    when the force ceased to act. What is this all about?

5. A particle has $\{E = 5\text{Kg}, p_x = 2\text{Kg}, p_y = 1\text{Kg}, p_z = 0\}$. What is
    its mass?

6. What are the components in Kg of the 4-momentum 4-vector of a particle moving in the x-direction through space at $0.8c$ with a rest-mass of 3Kg?

7. What is the momenergy 4-vector of a particle moving in the x-direction whose kinetic energy is half its rest energy?

8. Two particles, $\{A, B\}$, collide. In a given reference frame, the particles have 4-momentum vectors [4 2 1 2] & [5 3 1 0] respectively. When the particles separate, particle A has 4-momentum [6 3 3 0]. What is the initial total 4-momentum of the system in the given reference frame? What is the 4-momentum of particle B in the given reference frame?

9. If the length of the 4-momentum vector of a particle in a stationary observer's view is 4 Kg, what is the length of the 4-momentum vector of the particle in the view of an observer moving at half the velocity of light?

10. When it is stationary, the energy of a vice-chancellor's wallet is a million metric tons. When the wallet is moving at a velocity corresponding to the space-time angle χ, what is its spatial momentum and what is its energy?

18

DOING IT WITH MATRICES

We are now going to do in 2-dimensions, using the 2×2 space-time matrices, what we did in the previous two chapters with 4-vectors. The reader might have noticed that the 4-vector mathematics effectively carried two space dimensions along for the ride and that the essence of it was in just two of the components of a 4-vector. Within this chapter, we gain insights into the nature of velocity and acceleration in space-time.

In the standard presentation of 4-vectors above, we started with the position 4-vector $\overline{\overline{R}} = [x \quad y \quad z \quad ct]$ and differentiated this vector with respect to τ to get the 4-velocity.

After a little calculation, we came to:

$$\overline{\overline{U}} = \gamma(v)\begin{bmatrix} \vec{v} & c \end{bmatrix} \tag{18.1}$$

In 2-dimensions, we see that this corresponds to:

$$\begin{bmatrix} c\gamma & v\gamma \\ v\gamma & c\gamma \end{bmatrix} = \begin{bmatrix} \gamma & 0 \\ 0 & \gamma \end{bmatrix}\begin{bmatrix} c & v \\ v & c \end{bmatrix} \equiv \overline{\overline{U}} \equiv \begin{bmatrix} \cosh\chi & \sinh\chi \\ \sinh\chi & \cosh\chi \end{bmatrix} \tag{18.2}$$

Note that in units in which $c = 1$, $v < 1$. The velocity vector is the rotation matrix. The different magnitudes of velocity correspond to the different values of the space-time angle, χ. This really ought to be emphasized.

The velocity vector is the rotation matrix.
A velocity is just a space-time angle.

In the very first chapter of this book, the reader was told, without any discussion or justification, that change of velocity is just rotation in space-time. It was stated as if it were almost a "God-given truth". Now we see the basis of that "God-given" truth. A change in the angle in the rotation matrix is a change in velocity. Change of velocity is rotation (in space-time).

We have above (15.2):

$$\overline{\overline{U}} \bullet \overline{\overline{U}} = c^2 \tag{18.3}$$

This might have surprised the reader when she first saw it. We are doing no more than dotting together two rotation matrices (with the same angle, χ). Dotting the matrices together, we get:

$$\begin{bmatrix} \gamma & 0 \\ 0 & \gamma \end{bmatrix}\begin{bmatrix} c & v \\ v & c \end{bmatrix} \bullet \begin{bmatrix} \gamma & 0 \\ 0 & \gamma \end{bmatrix}\begin{bmatrix} c & v \\ v & c \end{bmatrix} = \begin{bmatrix} c^2 & 0 \\ 0 & c^2 \end{bmatrix} \tag{18.4}$$

This is:

$$\begin{aligned}
&\begin{bmatrix} \cosh \chi & \sinh \chi \\ \sinh \chi & \cosh \chi \end{bmatrix} \bullet \begin{bmatrix} \cosh \chi & \sinh \chi \\ \sinh \chi & \cosh \chi \end{bmatrix} \\
&= \begin{bmatrix} \cosh \chi & \sinh \chi \\ \sinh \chi & \cosh \chi \end{bmatrix}\begin{bmatrix} \cosh \chi & -\sinh \chi \\ -\sinh \chi & \cosh \chi \end{bmatrix} \\
&= \begin{bmatrix} \cosh^2 \chi - \sinh^2 \chi & 0 \\ 0 & \cosh^2 \chi - \sinh^2 \chi \end{bmatrix} \\
&\qquad\qquad\qquad = \begin{bmatrix} 1 & 0 \\ 0 & 1 \end{bmatrix}
\end{aligned} \tag{18.5}$$

$\overline{\overline{U}} \bullet \overline{\overline{U}} = c^2$ is no more than a simple trigonometric identity. It just seems profound in the 4-vector presentation.

If we do this in Euclidean space, we get the same:

$$\begin{bmatrix} \sigma & 0 \\ 0 & \sigma \end{bmatrix}\begin{bmatrix} 1 & g \\ -g & 1 \end{bmatrix} \bullet \begin{bmatrix} \sigma & 0 \\ 0 & \sigma \end{bmatrix}\begin{bmatrix} 1 & g \\ -g & 1 \end{bmatrix} = \begin{bmatrix} 1 & 0 \\ 0 & 1 \end{bmatrix} \tag{18.6}$$

18.1 DIFFERENTIATING THE VELOCITY VECTOR

In the 4-vector presentation, we differentiate the velocity 4-vector with respect to the invariant interval, τ, to get the acceleration

4-vector. τ is the square root of the determinant of the position vector matrix. In the space-time algebra, $\mathbb{S}$, we take the differential by differentiating with respect to both the co-ordinates (t, z). Doing this with matrices automatically takes account of the Cauchy Riemann equations of the $\mathbb{S}$ algebra. The reader is reminded that the Cauchy Riemann equations of a complex number system need to be satisfied by a function for the function to be differentiable.

We have:

$$\frac{\partial \begin{bmatrix} \cosh \chi & \sinh \chi \\ \sinh \chi & \cosh \chi \end{bmatrix}}{\partial \begin{bmatrix} t & z \\ z & t \end{bmatrix}}$$

$$= \begin{bmatrix} \sinh \chi \dfrac{\partial \chi}{\partial t} + \cosh \chi \dfrac{\partial \chi}{\partial z} & \cosh \chi \dfrac{\partial \chi}{\partial t} + \sinh \chi \dfrac{\partial \chi}{\partial z} \\ \cosh \chi \dfrac{\partial \chi}{\partial t} + \sinh \chi \dfrac{\partial \chi}{\partial z} & \sinh \chi \dfrac{\partial \chi}{\partial t} + \cosh \chi \dfrac{\partial \chi}{\partial z} \end{bmatrix}$$

(18.7)

This is the acceleration matrix. We call it the *Acceleration* vector to distinguish it from the 4-acceleration, $\overline{\overline{A}}$. Calculation shows that the determinant of this matrix, which is the magnitude of the acceleration vector, is:

$$\det(accel) = \left(\frac{\partial \chi}{\partial z} \right)^2 - \left(\frac{\partial \chi}{\partial t} \right)^2 = \left(\frac{\partial \chi}{\partial z} \right)^2 - \left(\frac{\partial \chi}{\partial z} \frac{\partial z}{\partial t} \right)^2$$

$$= \left(\frac{\partial \chi}{\partial z} \right)^2 (1 - v^2) = \gamma^{-2} \left(\frac{\partial \chi}{\partial z} \right)^2$$

(18.8)

This is positive for velocities less than the speed of light and reduces to zero at the limiting velocity – as it should do. On the time axis (stationary observer), $\chi = 0$ and the acceleration vector is:

$$\begin{bmatrix} \dfrac{\partial \chi}{\partial z} & \dfrac{\partial \chi}{\partial t} \\ \dfrac{\partial \chi}{\partial t} & \dfrac{\partial \chi}{\partial z} \end{bmatrix}$$

(18.9)

Normally, a stationary observer measures acceleration with respect to only time. It can be measured with respect to space (distance travelled) just as easily, but it normally is not so measured. This prejudice of ignoring of the rate of change of velocity with respect to space effectively sets $\dfrac{\partial \chi}{\partial z} = 0$ in our view. Of course, it is the spirit of special relativity that space and time are equivalent, and we ought to measure acceleration with respect to both space and time, as the matrix above does.

Looking at the above acceleration vector (matrix), we see that the rate of change of angle with respect to time is in the space direction (the off diagonal elements of the matrix). Of course, for a stationary observer, velocity is in the time direction. This is what corresponds to the 4-acceleration (time-acceleration) being orthogonal to the 4-velocity.

It is not that the space-time acceleration vector is at 90° to the space-time velocity vector, and thus orthogonal to that vector. Such a concept is nonsense in space-time. It is that the "time component" of acceleration is purely spatial and therefore will not lengthen the velocity vector in the time direction.

Notice how the acceleration co-mingles $\left(\dfrac{\partial v}{\partial t}, \dfrac{\partial v}{\partial z}\right)$. The acceleration vector is the whole matrix, and so the acceleration has two components.

We will need:

$$\frac{d\gamma}{dt} = \gamma^3 v \frac{dv}{dt} \tag{18.10}$$

$$\frac{d\gamma}{dz} = \gamma^3 v \frac{dv}{dz} \tag{18.11}$$

$$\frac{d(\gamma v)}{dt} = \gamma^3 v^2 \frac{dv}{dt} + \gamma \frac{dv}{dt} \tag{18.12}$$

$$\frac{d(\gamma v)}{dz} = \gamma^3 v^2 \frac{dv}{dz} + \gamma \frac{dv}{dz} \tag{18.13}$$

Now, we differentiate the velocity vector again:

$$\frac{d\begin{bmatrix} \gamma & v\gamma \\ v\gamma & \gamma \end{bmatrix}}{d\begin{bmatrix} t & z \\ z & t \end{bmatrix}} = \begin{bmatrix} \dfrac{d\gamma}{dt} + \dfrac{d(v\gamma)}{dz} & \dfrac{d\gamma}{dz} + \dfrac{d(v\gamma)}{dt} \\ \dfrac{d\gamma}{dz} + \dfrac{d(v\gamma)}{dt} & \dfrac{d\gamma}{dt} + \dfrac{d(v\gamma)}{dz} \end{bmatrix} \tag{18.14}$$

Substituting gives:

$$Acceleration = \begin{bmatrix} \gamma^3 v \dfrac{dv}{dt} + \gamma^3 v^2 \dfrac{dv}{dz} + \gamma \dfrac{dv}{dz} & \gamma^3 v \dfrac{dv}{dz} + \gamma^3 v^2 \dfrac{dv}{dt} + \gamma \dfrac{dv}{dt} \\ \gamma^3 v \dfrac{dv}{dz} + \gamma^3 v^2 \dfrac{dv}{dt} + \gamma \dfrac{dv}{dt} & \gamma^3 v \dfrac{dv}{dt} + \gamma^3 v^2 \dfrac{dv}{dz} + \gamma \dfrac{dv}{dz} \end{bmatrix} \quad (18.15)$$

In the previous chapter, we differentiated the dot product of the 4-velocity with itself to show that the 4-acceleration is orthogonal to the 4-velocity see. Unfortunately, the interpretation of the mathematics was in error because the cosh( ) function cannot be zero. We will now check the asserted orthogonality directly using the spacetime algebra of hyperbolic complex numbers. We will do this by using the inner product to calculate the space-time angle between the space-time velocity and the space-time acceleration. That spacetime angle is given by:

$$Acceleration \bullet velocity = \gamma^2 \begin{bmatrix} \dfrac{dv}{dz} & \dfrac{dv}{dt} \\ \dfrac{dv}{dt} & \dfrac{dv}{dz} \end{bmatrix} \quad (18.16)$$

The inner product is the real part of this and we have $\cosh \chi = \gamma^2 \dfrac{dv}{dz} \neq 0$.

For the stationary observer, if we are to be prejudiced, $\dfrac{dv}{dz} = 0$ and

we have that the inner product appears to be zero for the prejudiced

stationary observer. This is the origin of the alleged orthogonality of

the 4-acceleration, $\overline{\overline{\Lambda}}$, and the 4-velocity, $\overline{\overline{U}}$.

If we allow our prejudice and set $\dfrac{dv}{dz} = 0$ in the above acceleration, we get the prejudiced stationary observers view:

$$Acceleration \left(\dfrac{dv}{dz} = 0 \right) = \begin{bmatrix} \gamma^3 v \dfrac{dv}{dt} & \gamma^3 v^2 \dfrac{dv}{dt} + \gamma \dfrac{dv}{dt} \\ \gamma^3 v^2 \dfrac{dv}{dt} + \gamma \dfrac{dv}{dt} & \gamma^3 v \dfrac{dv}{dt} \end{bmatrix} \quad (18.17)$$

the algebra requires that:

$$\gamma^3 v \dfrac{dv}{dt} > \gamma^3 v^2 \dfrac{dv}{dt} + \gamma \dfrac{dv}{dt} : v < 1 \quad (18.18)$$

Which is not true unless $\dfrac{dv}{dt} < 0$. This is why we get the negative aspect of acceleration in 4-vector calculations. This compares with the 4-vector form:

$$\overline{\overline{A}} = \left[\gamma^4 \frac{(\vec{v} \bullet \vec{a})}{c^2} \vec{v} + \gamma^2 \vec{a} \quad \gamma^4 \frac{(\vec{v} \bullet \vec{a})}{c^2} c \right] \tag{18.19}$$

We have a factor of γ difference. This arises because we are differentiating matrices with respect to time while 4-vectors are differentiated with respect to τ - remember:

$$\frac{d\overline{\overline{U}}}{d\tau} = \gamma \frac{d\overline{\overline{U}}}{dt} \tag{18.20}$$

We are now going to dot the *Acceleration* vector with itself to calculate its norm:

$$\begin{bmatrix} \dfrac{d\gamma}{dt} + \dfrac{d(v\gamma)}{dz} & \dfrac{d\gamma}{dz} + \dfrac{d(v\gamma)}{dt} \\ \dfrac{d\gamma}{dz} + \dfrac{d(v\gamma)}{dt} & \dfrac{d\gamma}{dt} + \dfrac{d(v\gamma)}{dz} \end{bmatrix} \bullet \begin{bmatrix} \dfrac{d\gamma}{dt} + \dfrac{d(v\gamma)}{dz} & \dfrac{d\gamma}{dz} + \dfrac{d(v\gamma)}{dt} \\ \dfrac{d\gamma}{dz} + \dfrac{d(v\gamma)}{dt} & \dfrac{d\gamma}{dt} + \dfrac{d(v\gamma)}{dz} \end{bmatrix} \tag{18.21}$$

Multiplying out gives:

$$\begin{bmatrix} \left(\dfrac{d\gamma}{dt} + \dfrac{d(v\gamma)}{dz} \right)^2 - \left(\dfrac{d\gamma}{dz} + \dfrac{d(v\gamma)}{dt} \right)^2 & 0 \\ 0 & \sim \end{bmatrix} \tag{18.22}$$

Wherein, we have used "~" to avoid cluttering the page with duplicate information. This looks like an inner product should look. Substitution and calculation gives:

$$Acceleration \bullet Acceleration = \begin{bmatrix} \gamma^2 a'^2 \left(\dfrac{1}{v^2} \right) & 0 \\ 0 & \gamma^2 a'^2 \left(\dfrac{1}{v^2} \right) \end{bmatrix} \tag{18.23}$$

This seems very different from the answer we got in the standard presentation, which is:

$$\overline{\overline{A}} \bullet \overline{\overline{A}} \equiv \begin{bmatrix} -\gamma^6 a'^2 & 0 \\ 0 & -\gamma^6 a'^2 \end{bmatrix} \tag{18.24}$$

But let us redo the calculation to get the norm of the *Acceleration* vector as:

$$Acceleration \bullet Acceleration = \left[\left(\gamma^{4\ 2} + \gamma^2\right)\left(\left(\frac{dv}{dz}\right)^2 - \left(\frac{dv}{dt}\right)^{2'}\right) \atop 0 \right]_{\sim} \quad (18.25)$$

Let us be prejudiced and set $\dfrac{dv}{dz} = 0$ to get:

$$\left[\gamma^2\left(\gamma^2 v^2 + 1\right)\left(-\left(\frac{dv}{dt}\right)^2\right) \quad 0 \atop 0 \right]_{\sim} = \left[-\gamma^4 a'^2 \quad 0 \atop 0 \quad -\gamma^4 a'^2 \right] = \gamma^{-2} \overline{\overline{A}} \bullet \overline{\overline{A}} \quad (18.26)$$

The difference of the γ^2 factor can be traced to differentiating (once for each of the two acceleration 4-vectors) with respect to time instead of differentiating with respect to the invariant interval, τ. Taking the square root, which in our case is now very simple to do, we get the length of the acceleration vector:

$$|Acceleration| = \gamma \frac{a'}{v} \quad (18.27)$$

In the standard 4-vector presentation, we arrived at the acceleration transformation: $a' = \dfrac{a_0}{\gamma^3}$. In the matrix presentation, we have to adapt our length calculation to give us what the stationary observer sees. The prejudiced stationary observer will take the $\dfrac{dv}{dz}$ term to be zero because he measures acceleration with respect to only time. The stationary observer will "drop" the length of the acceleration vector on to the time axis by multiplying it by $\cosh \chi = \gamma$:

$$\gamma Acceleration \bullet \gamma Acceleration = \gamma^2 \left[\gamma^4 a^2(-1) \quad 0 \atop 0 \quad \gamma^4 a^2(-1) \right] \quad (18.28)$$

This leads to:

$$a' = \frac{a_0}{\gamma^3} \quad (18.29)$$

The (prejudiced) stationary observer has come to the same transformation of acceleration as the 4-vector standard presentation.

We did avoid having to fudge the square root of the norm, and we did sort out the dot product equals zero problem, but we have done more than that. We are now doing our calculations in space-time in a way that treats time and space equally. This equality is the true co-mingling of space and time; it is missed using 4-vectors. The matrix form is not just more elegant, it also gives the complete answer and not just a part of that complete answer. The matrix form gives us the "God's eye view" of space-time.

18.2 MOMENTUM

We assume that every particle in the universe has a rest-mass, m_0. In the standard presentation, we defined the momentum 4-vector as:

$$\overline{\overline{P}} = m_0 \overline{\overline{U}} = m_0 \gamma(u)\begin{bmatrix} \vec{u} & c \end{bmatrix} \tag{18.30}$$

We similarly define the momentum matrix as the product of scalar mass matrix and the velocity matrix:

$$\begin{bmatrix} m_0 & 0 \\ 0 & m_0 \end{bmatrix}\begin{bmatrix} \gamma & v\gamma \\ v\gamma & \gamma \end{bmatrix} = \begin{bmatrix} m_0\gamma & m_0\gamma v \\ m_0\gamma v & m_0\gamma \end{bmatrix} \equiv \overline{\overline{P}} \tag{18.31}$$

We immediately have that one component of this is the energy, $m_0\gamma$, and the other component is the momentum, $m_0\gamma v$. Adjusting the units gives:

$$\begin{bmatrix} m_0\gamma & m_0\gamma v \\ m_0\gamma v & m_0\gamma \end{bmatrix} = \begin{bmatrix} \dfrac{E}{c} & p \\ p & \dfrac{E}{c} \end{bmatrix} \equiv \begin{bmatrix} m_0 \cosh\chi & m_0 \sinh\chi \\ m_0 \sinh\chi & m_0 \cosh\chi \end{bmatrix} \tag{18.32}$$

We get the unification that we found using 4-vectors. Note that the mass-dimensions of all elements of the matrix are the same. We also have that the ratio of the energy to the spatial momentum is just the ratio of the cosh( ) function to the sinh( ) function. We could set the units of mass so that $m_0 = 1$, and we would thus have:

$$\begin{aligned} Energy &= \cosh\chi \\ Momentum &= \sinh\chi \end{aligned} \tag{18.33}$$

What is the scalar mass matrix? We are working in a division algebra (the hyperbolic complex numbers); the scalar mass matrix must be an element of this algebra; it must be of the form:

$$\begin{bmatrix} m_0 & 0 \\ 0 & m_0 \end{bmatrix} = \begin{bmatrix} \cosh \chi & \sinh \chi \\ \sinh \chi & \cosh \chi \end{bmatrix} : \chi = 0 \qquad (18.34)$$

This is why everything in space-time has a non-zero rest energy (rest-mass). It is because the minimum of the cosh( ) function is unity and not zero. That's worth emboldening.

Everything has a non-zero rest energy (rest-mass)
because the minimum of the cosh( ) function is unity.

Hang on a minute! We spoke earlier, about humankind being stuck with the radial component of the hyperbolic complex numbers being set at $h = 1$. If the radial component is so stuck at unity, then it makes sense to ascribe the non-zero rest mass of a particle to the non-zero minimum of the cosh( ) function, but the scalar mass could be of the form of the radial component of the algebra. Perhaps that is what mass is, but then we have to ask why different masses?

The Lorentz transformation of the energy-momentum vector is:

$$\begin{bmatrix} \gamma & v\gamma \\ v\gamma & \gamma \end{bmatrix} \begin{bmatrix} E & p \\ p & E \end{bmatrix} = \begin{bmatrix} \gamma(E+vp) & \gamma(p+vE) \\ \gamma(p+vE) & \gamma(E+vp) \end{bmatrix}$$

$$\begin{bmatrix} \cosh \chi & \sinh \chi \\ \sinh \chi & \cosh \chi \end{bmatrix} \begin{bmatrix} E & p \\ p & E \end{bmatrix} \qquad (18.35)$$

$$= \begin{bmatrix} E\cosh \chi + p\sinh \chi & p\cosh \chi + E\sinh \chi \\ p\cosh \chi + E\sinh \chi & E\cosh \chi + p\sinh \chi \end{bmatrix}$$

Conservation of momenergy is the invariance of the determinant under rotation (the rotation matrix has determinant unity, of course). Taking the determinants of both sides of the above will show:

$$\det\left(\begin{bmatrix} E & p \\ p & E \end{bmatrix} \right) = \det\left(\begin{bmatrix} \gamma(E+vp) & \gamma(p+vE) \\ \gamma(p+vE) & \gamma(E+vp) \end{bmatrix} \right) \qquad (18.36)$$

Momenergy is the determinant of the energy-momentum vector, and conservation of momenergy is no more than invariance of the determinant under rotation in space-time.

It is important to realize that the conservation of momenergy law breaks into the conservation of energy law and the conservation of momentum law for a stationary observer. This means that a stationary observer will always see two conservation laws (energy and momentum). This is like space-time breaks into space and time for a stationary observer and electromagnetic force breaks into electric force and magnetic force for a stationary observer.

18.3 FORCE

In the standard 4-vector presentation, we defined force as the 4-vector:

$$\bar{\bar{F}} = \frac{d\bar{\bar{P}}}{d\tau} \tag{18.37}$$

However, using matrices, we take the complete differential rather than differentiate with respect to the invariant interval, τ. The differential of the momentum matrix is:

$$d\begin{bmatrix} m_0\gamma & m_0\gamma v \\ m_0\gamma v & m_0\gamma \end{bmatrix} = \begin{bmatrix} m_0\left(v\dfrac{d\gamma}{dz} + \gamma\dfrac{dv}{dz} + \dfrac{d\gamma}{dt}\right) + \gamma v\dfrac{dm_0}{dz} + \gamma\dfrac{dm_0}{dt} & \sim \\ m_0\left(\dfrac{d\gamma}{dz} + v\dfrac{d\gamma}{dt} + \gamma\dfrac{dv}{dt}\right) + \gamma\dfrac{dm_0}{dz} + \gamma v\dfrac{dm_0}{dt} & \sim \end{bmatrix} \tag{18.38}$$

Substituting and putting $\dfrac{dm_0}{dz} = \dfrac{dv}{dz} = 0$ gives:

$$d\begin{bmatrix} m_0\gamma & m_0\gamma v \\ m_0\gamma v & m_0\gamma \end{bmatrix} = \begin{bmatrix} m_0\gamma^3 v\dfrac{dv}{dt} + \gamma\dfrac{dm_0}{dt} & \sim \\ m_0\gamma^3 v^2\dfrac{dv}{dt} + m_0\gamma\dfrac{dv}{dt} + \gamma v\dfrac{dm_0}{dt} & \sim \end{bmatrix} \tag{18.39}$$

In the standard presentation, we had:

$$\bar{\bar{F}} = \frac{d\bar{\bar{P}}}{d\tau} = \gamma\left[m_0\gamma^3 v\frac{dv}{dt}\vec{v} + m_0\gamma\vec{a} + \gamma v\frac{dm_0}{dt} \quad \left(m_0\gamma^3 v\frac{dv}{dt} + \gamma\frac{dm_0}{dt}\right)c\right] \tag{18.40}$$

Thus, other than units, we agree with the standard presentation when $\dfrac{dm_0}{dz} = \dfrac{dv}{dz} = 0$ except for the factor of γ which, as in the acceleration vector, derives from differentiating with respect to the invariant interval.

Taking the differential of the energy momentum vector gives:

$$d\begin{bmatrix} E & p \\ p & E \end{bmatrix} = \begin{bmatrix} \dfrac{dE}{dt} + \dfrac{dp}{dz} & \dfrac{dp}{dt} + \dfrac{dE}{dz} \\ \dfrac{dp}{dt} + \dfrac{dE}{dz} & \dfrac{dE}{dt} + \dfrac{dp}{dz} \end{bmatrix} \qquad (18.41)$$

In the standard 4-vector presentation, we also have:

$$\overline{\overline{F}} = \frac{d\overline{\overline{P}}}{d\tau} = \gamma \begin{bmatrix} \vec{F} & \dfrac{1}{c} power \end{bmatrix} \qquad (18.42)$$

As seen by the prejudiced stationary observer, $\dfrac{dp}{dz} = \dfrac{dE}{dz} = 0$ and the γ factor is from the differentiation, but the real situation is that we have:

$$Power = \frac{dE}{dt} + \frac{dp}{dz}$$
$$Force = \frac{dp}{dt} + \frac{dE}{dz} \qquad (18.43)$$

Because of the cosh( ) > sinh( ) nature of space-time, we have *Power* > *Force*. Such is the symmetry of special relativity's understanding of the equivalence of space and time.

18.4 A LITTLE FOOD FOR THOUGHT

We remind the reader that a 2-component vector field $\{u(t, z), v(t, z)\}$ over space-time has:

$$\begin{bmatrix} Div(F) & Curl(F) \\ Curl(F) & Div(F) \end{bmatrix} = \begin{bmatrix} \dfrac{du}{dt} + \dfrac{dv}{dz} & \dfrac{dv}{dt} + \dfrac{du}{dz} \\ \dfrac{dv}{dt} + \dfrac{du}{dz} & \dfrac{du}{dt} + \dfrac{dv}{dz} \end{bmatrix} \qquad (18.44)$$

$$: \ Div(F) > Curl(F)$$

We see that, if energy and momentum are thought of as the two components of a vector field over space-time, then power is the divergence of that vector field and force is the curl of that vector field.

The standard presentation of special relativity mechanics is to use 4-vectors, but we have shown that special relativity mechanics can be presented using the hyperbolic complex numbers of space-time. The hyperbolic complex numbers are a proper algebraic structure (they are an algebraic field) whereas the 4-vectors are not a proper algebraic structure (they are not a division algebra). The hyperbolic complex numbers automatically have a finite speed (the speed of light) as part of their structure, and so we know that there is nothing outside of the algebra (faster than light space-time). Without the hyperbolic complex numbers, special relativity gets the correct results, but it is somewhat contrived, whereas, with the hyperbolic complex numbers, everything just falls out of the finite group C_2.

The hyperbolic complex numbers is a 2-dimensional algebra. Special relativity is 2-dimensional, and it is only the cosmetic inclusion of two inert spatial dimensions that makes special relativity appear to be 4-dimensional. We observe a 4-dimensional space-time in the universe. There seems to be a difficulty here. The difficulty is resolved by considering the Lorentz group, and we will deal with this in due course. It is, of course, the choice of the reader as to whether he prefers the hyperbolic complex numbers of the 4-vector algebra.

18.5 FIVE FUNDAMENTAL VECTORS AGAIN

The reader might recall that in the chapter on vectors, we listed five fundamental vectors, displacement, velocity, momentum (velocity multiplied by mass), acceleration, force (acceleration multiplied by mass). Perhaps, we ought to have mentioned the electromagnetic field vector as well. What are these five fundamental vectors in space-time?

The velocity vector is the time and space vector given by:

$$velocity \equiv \begin{bmatrix} t & z \\ z & t \end{bmatrix} \equiv \begin{bmatrix} \cosh \chi & \sinh \chi \\ \sinh \chi & \cosh \chi \end{bmatrix} \tag{18.45}$$

The acceleration vector is given by

$$acceleration \equiv \begin{bmatrix} \dfrac{\partial \chi}{\partial z} & \dfrac{\partial \chi}{\partial t} \\[2mm] \dfrac{\partial \chi}{\partial t} & \dfrac{\partial \chi}{\partial z} \end{bmatrix} \equiv \begin{bmatrix} \cosh \kappa & \sinh \kappa \\ \sinh \kappa & \cosh \kappa \end{bmatrix} \qquad (18.46)$$

The momentum vector is given by:

$$Momentum \equiv \begin{bmatrix} E & p \\ p & E \end{bmatrix} \equiv m_0 \begin{bmatrix} \cosh \chi & \sinh \chi \\ \sinh \chi & \cosh \chi \end{bmatrix} \qquad (18.47)$$

The force vector is given by:

$$Force \equiv \begin{bmatrix} Power & Force_{Spatial} \\ Force_{Spatial} & Power \end{bmatrix} \equiv m_0 \begin{bmatrix} \cosh \kappa & \sinh \kappa \\ \sinh \kappa & \cosh \kappa \end{bmatrix} \qquad (18.48)$$

The electromagnetic vector is given by

$$E.Mag \equiv \begin{bmatrix} E & B \\ B & E \end{bmatrix} \equiv \begin{bmatrix} \cosh \eta & \sinh \eta \\ \sinh \eta & \cosh \eta \end{bmatrix} \qquad (18.49)$$

We are missing the displacement vector. What is displacement in time? It is age. We are thus led to the idea that the displacement vector is:

$$Displacement \equiv \begin{bmatrix} Age & Extent \\ Extent & Age \end{bmatrix} \equiv \begin{bmatrix} \cosh \rho & \sinh \rho \\ \sinh \rho & \cosh \rho \end{bmatrix} \qquad (18.50)$$

It might thereby seem that the age of the universe is given by the cosh( ) function and the extent of the universe is given by the sinh( ) function. If so, then as the age of the universe increases, the extent of the universe will increase – we call this the expanding universe. Of course, the cosh( ) function is never zero, and so the age of the universe was never zero; however, the sinh( ) function can be zero, and so the extent of the universe was once zero – there's a thought.

The point is that all five fundamental vectors, and the electromagnetic vector, can be drawn on "space-time" axes (with the asymptotes) such that the electric field takes the place of time and the magnetic field takes the place of space or power takes the place of time and force takes the place of space and so on. There we have it! The cosh( ) and sinh( ) functions have it all – thank you the exponential function.

19

ELECTROMAGNETISM

"What led me more or less directly to the theory of special relativity…(was the realisation that)…the electromagnetic force acting on a (charged) body in motion in a magnetic field was nothing else but an electric field (in the rest frame)"

Albert Einstein[1].

An electricity generator is comprised of a magnet (that produces a magnetic field) and a coil of (copper) wire. If we move the wire so that the moving wire cuts the magnetic field lines, an electric current is produced in the wire. If we equally move the magnet so that the wire cuts the moving magnetic field lines, an equal electric current is produced in the wire. Thus, this electromagnetic phenomenon is invariant under change of view from the stationary observer to the moving observer. This indicates that electromagnetism is Lorentz invariant, which means that electromagnetism and special relativity fit together perfectly. It could have been that, instead of being invariant under Lorentz transformations, electromagnetism was invariant under Newtonian transformations, but it isn't – the universe is not Newtonian.

Aside: The average speed of electrons in a wire carrying a 10 amp electrical current is approximately 1 millimeter per second.

[1.] Am. J. Phys. 32, 16 (1964) pg 35 R. S. Shankland.

Electric fields are determined by charge distribution. Magnetic fields are determined by current distribution. Electric charge moves through time, and so electric charge is a current in the time direction. If we drive a (spatial) electric current in the same direction through two parallel wires, the wires will attract each other. We explain this by saying that the current produces a magnetic field that circulates in the same direction around the wires. In between the two wires, the two magnetic fields of the two wires point in opposite directions. Since opposite magnetic poles attract, the opposite directions of the magnetic fields attract each other and the wires are attracted towards each other.

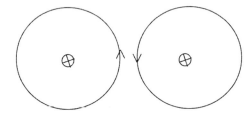

Aside: Why do opposite magnetic poles attract each other? It seems that the magnetic fields try to cancel each other and thereby reduce the energy in the system. There seems to be a "motivation" within the universe to seek out the lowest energy level. If the energy level will be reduced by two current carrying wires being close to each other, then the wires will attract each other until the energy of the tension in the wires that holds them apart matches the energy reduction due to the proximity of the wires. It seems that all attractive/repulsive forces work on this energy level reduction basis. Why does the universe try to lower the energy level of a system? – We don't know.

We have the 2-dimensional plane of the magnetic field. We have an electrical current perpendicular to that magnetic plane, and we have time. For the first time in this book, we have met a 4-dimensional phenomenon. Prior to this chapter, we were carrying two inert spatial dimensions for no apparent reason other than we felt naked without them. Clearly, we will not be able to explain the whole of electromagnetism with only 2-dimensional spaces – (we're going to need the quaternions). It turns out that the order four group

$C_2 \times C_2$ is involved; this group has three order two sub-groups and the 4-dimensional spaces in it have three 2-dimensional sub-spaces. Thus, to some extent, we will be able to present 2-dimensional aspects of electromagnetism.

If we switch off the current, the magnetic field disappears. The strength of the magnetic force depends on the amount of the current flowing through the wires. A higher current produces a stronger magnetic field. A higher current, through the same wire, is faster moving electrons or more electrons moving at the same speed. The magnetic force depends on the amount and velocity of the electric charge through space, which is the current. Here, we have that force is dependent upon velocity. Force is not dependent upon velocity in Newtonian mechanics; Newtonian mechanics cannot apply to electromagnetism.

19.1   THE MAXWELL EQUATIONS I

In the standard presentation of physics, we are told that all electromagnetic phenomena are derivable from the Lorentz force law and the four equations known as the Maxwell equations. The Lorentz force law and the Maxwell equations are vector equations. The Maxwell equations were first written using quaternions rather than vectors by James Clerk Maxwell in 1861/62, but later re-written in vector form by Gibbs and Heaviside. It is not true that all electromagnetic phenomena are derivable from the four Maxwell equations and the Lorentz force law. The quantization of electric charge (the charge of the electron), the nature of electron orbits in atoms, the wave-particle nature of electrons, and the electro-weak unification are, at least, some of the electromagnetic phenomena that are not derivable from the Maxwell equations and the Lorentz force law – well! they have not been derived so far. None-the-less, all "classical" electromagnetism is expressed by the Maxwell equations and the Lorentz force law[2], and a course covering the Maxwell equations and the Lorentz force law is considered a complete course in classical electromagnetism.

[2.] Perhaps we ought to include the continuity equation and the equation defining the field.

Maxwell was led to the equations named after him by an inconsistency in Ampere's[3] law. Ampere's law is:

$$Curl(\vec{B}) = \vec{j} \qquad (19.1)$$

Taking the divergence gives:

$$Div(Curl(\vec{B})) = 0 = Div(\vec{j}) \qquad (19.2)$$

This cannot be true. The divergence of any curl is always zero, but the divergence of a current is not zero. Maxwell corrected Ampere's law by adding $\dfrac{\partial \vec{E}}{\partial t}$ to the right-hand side of the equation, and was thus led to the Maxwell equations. The $\dfrac{\partial \vec{E}}{\partial t}$ bit is electromagnetic waves; these were detected by Heinrich Hertz in 1888 thereby asserting the correctness of the Maxwell equations.

After Maxwell had produced his equations, it was noticed by physicists that the equations are invariant under Lorentz transformations (rotation in space-time), but that they were not invariant under the traditional Galilean transformations associated with the Newtonian view of space and time. It is because the Maxwell equations are Lorentz invariant that the electrical permittivity, ε_0, and the magnetic permeability, μ_0, are Lorentz invariant and thus the speed of light, given by $c = \dfrac{1}{\sqrt{\varepsilon_0 \mu_0}}$, is Lorentz invariant. Actually, it is the other way around. Because all bodies travel through space-time at the same speed (the speed of light)[4], both $\{\varepsilon_0, \mu_0\}$ have to be Lorentz invariant and so any correctly written equations containing them will have to be Lorentz invariant.

It has been shown by Purcell[5] that, provided the laws of physics are Lorentz invariant (the same at every velocity) and electromagnetic attraction is an inverse square law (Coulomb's law) and electric charge is conserved, then magnetic effects are bound to happen

---

[3.] Ampere (1775–1836).
[4.] Which is because it is the nature of space-time that 4-acceleration is "orthogonal" to 4-velocity in space-time and thus 4-velocity can change in direction but not in magnitude.
[5.] Edgar M. Purcell, Electricity and Magnetism, New York 1963 pg. 173.

and the Maxwell equations can be derived from no more than these premises. So, given space-time and a small number of other reasonable assumptions, we must have electromagnetism. Intriguingly, we will see in due course that Maxwell's equations fall out of quaternion space.

In the standard presentation, the equations of electromagnetism are:

i. The two scalar (just real numbers on either side of the equals sign) Maxwell Equations:

$$Div\left(\vec{E}\right) = \frac{\partial \vec{E_x}}{\partial x} + \frac{\partial \vec{E_y}}{\partial y} + \frac{\partial \vec{E_z}}{\partial z} = \frac{\rho}{\varepsilon_0}$$

$$Div\left(\vec{B}\right) = \frac{\partial \vec{B_x}}{\partial x} + \frac{\partial \vec{B_y}}{\partial y} + \frac{\partial \vec{B_z}}{\partial z} = 0$$

(19.3)

Wherein, $\vec{E}$ is the electric field vector, $\vec{B}$ is the magnetic field vector, ρ is the electric charge density (Coulomb M^{-3})[6], and ε_0 is the electrical permittivity of empty space ($C^2 N^{-1} M^{-2}$).

Aside: Two charges of one Coulomb each would exert a force upon each other of one Newton if they were separated by 100 kilometers – so a coulomb is quite a large amount of electrical charge, or the electromagnetic force is very strong.

---

ii. The two vector (vectors on either side of the equals sign) Maxwell equations are:

$$Curl\left(\vec{E}\right) = -\frac{\partial \vec{B}}{\partial t}$$

$$Curl\left(\vec{B}\right) = \mu_0 \vec{j} + \mu_0 \varepsilon_0 \frac{\partial \vec{E}}{\partial t}$$

(19.4)

These equations say that, if the fields are not changing with time, then the electric field is an irrotational field but the

---

[6.] Augustin de Coulomb (1736–1806). In 1785, he performed systematic and precise experiments that discovered the law of the force between charged bodies known as Coulomb's Law.

magnetic field is a rotational field - that's rotation in a 2-dimensional Euclidean plane.

Aside: The Lie group $U(1)$ is rotation in the 2-dimensional Euclidean complex plane. Thus, we ought to expect that this Lie group will appear somewhere in electromagnetism, and it does.

Written in component form, these are:

$$
\begin{bmatrix}
\dfrac{\partial \vec{E}_z}{\partial y} - \dfrac{\partial \vec{E}_y}{\partial z} \\[2ex]
\dfrac{\partial \vec{E}_x}{\partial z} - \dfrac{\partial \vec{E}_z}{\partial x} \\[2ex]
\dfrac{\partial \vec{E}_y}{\partial x} - \dfrac{\partial \vec{E}_x}{\partial y}
\end{bmatrix}
=
\begin{bmatrix}
-\dfrac{\partial \vec{B}_x}{\partial t} \\[2ex]
-\dfrac{\partial \vec{B}_y}{\partial t} \\[2ex]
-\dfrac{\partial \vec{B}_z}{\partial t}
\end{bmatrix}
\tag{19.5}
$$

$$
\begin{bmatrix}
\dfrac{\partial \vec{B}_z}{\partial y} - \dfrac{\partial \vec{B}_y}{\partial z} \\[2ex]
\dfrac{\partial \vec{B}_x}{\partial z} - \dfrac{\partial \vec{B}_z}{\partial x} \\[2ex]
\dfrac{\partial \vec{B}_y}{\partial x} - \dfrac{\partial \vec{B}_x}{\partial y}
\end{bmatrix}
= \mu_0
\begin{bmatrix}
j_x + \varepsilon_0 \dfrac{\partial \vec{E}_x}{\partial t} \\[2ex]
j_y + \varepsilon_0 \dfrac{\partial \vec{E}_y}{\partial t} \\[2ex]
j_z + \varepsilon_0 \dfrac{\partial \vec{E}_z}{\partial t}
\end{bmatrix}
\tag{19.6}
$$

Wherein, $\vec{j}$ is the electrical current density (the amount of electrical charge per second passing perpendicularly through a square meter) $(CS^{-1}M^{-2})$; μ_0 is the magnetic permeability of empty space (*Newton* $S^2 C^{-2}$). Note that the current in these equations flows from positive to negative not in the direction of the electron velocity. There are, effectively, six scalar equations here, and so we can think of Maxwell's equations as being eight scalar equations, but the six (vector) equations are entwined together in a way expressed by the vector notation. That entwinement is the fact that a vector is the same length in any co-ordinate system whereas the components of the vector in the three scalar equations change from one co-ordinate system to another; the entwinement is that the components changing under rotation compensate for each other to keep the vector's length constant. We do not want to miss that entwinement for our understanding would be less without it; it would be better if we could entwine all eight equations together.

Nothing in the Maxwell equations describes the effect of the fields on charged particles. For that, we need:

iii. The (vector) Lorentz force law is:

$$\vec{F} = q\vec{E} + q\vec{v} \times \vec{B} \tag{19.7}$$

wherein q is the electric charge of a body and $\vec{v}$ is the velocity through space (relative to the observer) of that charged body. In component form, this is:

$$\begin{bmatrix} F_x \\ F_y \\ F_z \end{bmatrix} = q \left(\begin{bmatrix} E_x \\ E_y \\ E_z \end{bmatrix} + \begin{bmatrix} v_y B_z & v_z B_y \\ v_z B_x - v_x B_z \\ v_x B_y & v_y B_x \end{bmatrix} \right) \tag{19.8}$$

It is drawn to the reader's attention that the force in this equation depends upon the velocity of the charged body; this equation can never fit into a Newtonian universe where force is invariant under change of velocity. If the velocity through space is zero, we have that the force is due to only the electric charge (Coulomb's law). If the velocity through space is not zero, we have two components of the force - electrical force and magnetic force.

The velocity is in a 3-dimensional vector cross product with the magnetic field vector, and so, if the velocity is at an angle of 0° to the magnetic field, we will still have force due to only the electric field. The force depends, not only upon the magnitude of the velocity, but also upon its direction relative to the magnetic field. We could not annul the electric force be changing the space direction in which we pointed a charged body, but we can annul the magnetic field by changing the spatial direction of the observer's motion.

The reader's attention is drawn to the fact that a 2-dimensional Euclidean rotational field (the magnetic field) arises from a rotation in space-time (velocity of a charged body). To repeat, a space rotation phenomenon arises from a space-time rotation. What does this mean? We will see something like this when we look at the Lorentz group and find that the commutator of two boosts (rotations in space-time) is a rotation in space.

For completeness, we also give the scalar continuity equation, which expresses the conservation of electric charge and which is:

$$\frac{\partial \rho}{\partial t} + \frac{\partial \vec{J}_x}{\partial x} + \frac{\partial \vec{J}_y}{\partial y} + \frac{\partial \vec{J}_z}{\partial z} = \frac{\partial \rho}{\partial t} + Div\left(\vec{j}\right) = 0 \qquad (19.9)$$

Does this not encourage us to think of electric charge density, ρ, as an electric current in the time direction? If we are to have charge conservation, the continuity equation must have the same form in all reference frames.

The definition of the electric field is:

$$\vec{E} = \frac{\vec{F}}{q} \qquad (19.10)$$

which is just the vector equation in which force is divided by scalar electric charge, q.

Electromagnetic fields can be superimposed on to other electromagnetic fields by simply adding the field vectors at each point in space. This is called superposition, and it is an important, and simplifying, property of electromagnetism. It is due to the linearity of the electromagnetic equations. Such linearity makes you think that perhaps the equations ought to be written as matrices.

Aside: The ratio of the gravitational force between two electrons to the electromagnetic force between them is $\approx 10^{42}$

19.2   THE INVARIANCE OF ELECTRIC CHARGE

To a stationary observer, the electrical charge of an electron is the same in all reference frames – it does not care about directions in space-time; it is a physical constant. Electrical charge is invariant under change of velocity; the amount of it does not dilate of contract with change of velocity. Atoms are comprised of a positively charged nucleus and negatively charged electrons that orbit the nucleus. Atoms are (when not ionized) electrically neutral as we can easily observe in the macroscopic world. Because electrons and protons move at different and varying speeds[7] in and around the nucleus, atoms can be consistently electrically neutral only if electric charge is the same at all velocities. Further, we observe that, an ionized atom does not become more or less ionized when moving than when stationary and rapidly moving electrons are not more or less charged than slowly moving electrons. When a positron and an electron collide (head on is simplest, but any collision will do) and annihilate each other, they produce electrically neutral photons. Such an event can be seen from different reference frames. If electric charge varied with velocity, then, unless the electron and the positron were moving together, the resulting photons could not be electrically neutral or, in such cases, we would not have conservation of electric charge. We can take the view that conservation of electrical charge compels electrical charge to be invariant under change of velocity. It would be a most strange universe if electric charge was not invariant under change of velocity.

Why should electrical charge be a Lorentz invariant (does not change with velocity)? Mass is not a Lorentz invariant, although rest-mass is invariant. There is another difference between mass and electric charge. If we add a proton to a neutron, the resulting nucleus has a mass that is not equal to the sum of the masses of the two constituent particles because of the binding energy holding them together. If we put two electrons together, the resultant "nucleus" has an exactly equal amount of electric charge as the two constituent particles. We do not know why electrical charge is invariant under change of velocity. It is as if electrical phenomena are not part of space-time, or do not happen in space-time, or are not connected to space-time, or do not "feel" direction in space-time. We say that electrical charge does not couple to space-time.

Charge density, which is the amount of electrical charge in a given volume of space, is not Lorentz invariant. That's quite important; I think I'll say it again. Charge density is not Lorentz invariant. Because a stationary observer sees the length of a moving volume

---

[7.] We assume that electrons move at varying speeds around the nucleus; they do seem to be at different distances from the nucleus – Rydberg atoms.

contract in the direction of the motion, she sees the actual volume contract. For a moving volume, we have:

$$Volume = length \times hieght \times width$$
$$Volume' = l'.h'.w' = l'.h_0.w_0 \tag{19.11}$$
$$= \sqrt{1 - v^2}\, l_0.h_0.w_0 = Volume_0 \sqrt{1 - v^2}$$

Charge density is charge divided by volume, and so we see that a stationary observer will see the charge density of a moving volume to be greater than the charge density of the volume when stationary. Greater charge density means greater electromagnetic force per unit volume (or length).

19.3 MAGNETIC EFFECTS

Consider a stationary wire with no electrical charge but carrying a current. An electrically charged body moving parallel to the wire at velocity v, "feels" a magnetic force (attracting or repelling according to the direction of the current) from the wire. (A moving charged body is a current just as much as moving electrons are a current.) However, to an observer also moving at velocity v, the electrically charged body is stationary and therefore should not feel a magnetic force from the wire. None-the-less, the moving observer must see a force affecting the motion of the charged body. This force, as seen by the moving observer, is not magnetic; it is electrical – think Lorentz force law. The Lorentz force law is:

$$\vec{F} = q\vec{E} + q\vec{v} \times \vec{B} \tag{19.12}$$

Where $\vec{v}$ is the velocity through space (not space-time).

An observer who moves at a velocity different from the velocity of the charged body will see the magnetic part of this force because the relative velocity between them will not be zero. An observer who moves at the same velocity as the charged body will see:

$$\vec{F} = q\vec{E} \tag{19.13}$$

One observer's magnetic force is another observer's electrostatic force. This is like one observer's time is another observer's space-time. But we would expect that they both observe a force because they will both observe a deflection of the path through space-time (or just space) of the charged body.

The stationary observer sees a magnetic field of magnitude determined by the velocity of the charged body relative to her. The moving observer sees a magnetic field of magnitude that is determined by the velocity of the electric charge relative to him. Because observers in motion relative to each other see the electric charge moving at different velocities, they see different amounts of current, and, thus, see magnetic fields of different magnitudes. If one observer moves at the same velocity as the electric charge (the charged body or the electrons in the wire), then that observer sees zero magnetic field[8].

Aside: Magnetic monopoles are postulated to be quantitised amounts of plus or minus magnetism. They are the "mirror image" of electrons and positrons with "magnetic charge" in the place of electric charge. Although Dirac and others have predicted magnetic quantization and it has a place in superconductor theory, magnetic monopoles have never been observed. When we dealt with simultaneity in an earlier chapter, we found that there is always an observer moving at a particular velocity that will see two spatially separate but close events as having zero time separation; the events are separated by pure space; they are simultaneous. So, with a charged body, it is always possible to find a velocity at which the electromagnetic force is purely electrostatic. Now, the nature of time is that the cosh( ) function is never less than one, and so the time axis co-ordinate is never less than one. Since the sinh( ) function can equal zero, the space axis co-ordinate can be zero. If we identify the electric force with time and the magnetic force with space, then we might expect a non-zero amount of electric charge (the charge of the electron perhaps) and also expect a zero magnetic charge – no magnetic monopoles.

A stationary current carrying wire is a lot of positively charged ions fixed in a crystal matrix that is stationary and a lot of negatively charged moving electrons flowing along the wire. For the stationary observer, the stationary wire is electrically neutral (put a stationary

---

[8]. Remember that the direction of the velocity is important.

electric charge near it), and so the charge density of the positive ions must be equal to the charge density of the moving electrons. Think about that; the number of electrons per stationary meter length of wire has got to equal the number of protons per stationary meter length of the wire. But, meter length changes with velocity (length contraction), and so, to the stationary observer, the meter of electrons is shorter than the meter of protons. There can be an electrostatic balance only if the density of electrons in the moving meter is less than in the density of protons in the stationary meter.

This electrostatic imbalance is apparent to the observer moving with an electrically charged body because, in his view, the electrons are not moving and thus there are more protons per meter than electrons; this effect is exacerbated by the contraction of the proton meter as seen by the moving observer. Thus, the moving observer sees an electrostatic force affecting the, stationary in his view, charged body and the stationary observer sees a magnetic force affecting the, moving in her view, charged body. Yes! it is a little difficult to untangle.

If the above paragraph is difficult to accept, do the experiment. Place a stationary electrically charged body near to a current carrying wire; vary the current if you like – try zero current. You will find that the stationary electrically charged body does not feel any force from the current carrying wire. Of course, if the electrically charged body is not stationary (subject to the cross-product in the Lorentz force equation), it does feel a force from the current carrying wire (provided the current is not zero).

Aside: The dimensions of the electric field can be calculated from the definition of the electric field as:

$$E = \frac{F}{q} \equiv \frac{Newtons}{Coulombs} \Rightarrow$$

$$[E] = Newton^1 Coulomb^{-1}.$$

The dimensions of the magnetic field can be calculated from the Lorentz force law:

$$F = qE + qv \times B \sim B$$

$$= \frac{F}{qv} \Rightarrow [B]$$

$$= \text{Newton}^1 \text{Coulomb}^{-1} \text{Sec}^1 \text{Meter}^{-1}$$

The magnetic effects of which we have spoken are all a consequence of length contraction. Length contraction is a consequence of the nature of rotation in space-time and the fact that we can rotate in space-time (which we could not do if space and time were not commingled together). Space-time comes from the finite group C_2. One can look at C_2 as being the multiplicative relations between $\{-1, +1\}$, and so we might think that magnetism exists because the numbers $\{-1, +1\}$ exist. With a little thought, we realize that electromagnetism is a 4-dimensional phenomenon whereas C_2 produces only 2-dimensional phenomena. There must be more to electromagnetism than only $\{-1, +1\}$. In later chapters, we will see that the group $C_2 \times C_2$ produces 4-dimensional phenomena, and so, in that sense, magnetism exists because the numbers $\{-1, +1\}$ exist.

19.4 ELECTROMAGNETIC WAVES

One of the great triumphs of the Maxwell equations is that they predict electromagnetic waves which are identified with light (also gamma rays, x-rays, infrared, radio waves etc.). The Maxwell equations (with $c = 1$) are:

$$\nabla \bullet \vec{E} = \rho \qquad\qquad \nabla \bullet \vec{B} = 0$$

$$\nabla \times \vec{E} = -\frac{\partial \vec{B}}{\partial t} \qquad\qquad \nabla \times \vec{B} = \frac{\partial \vec{E}}{\partial t} + \vec{j} \qquad (19.14)$$

Wherein $\nabla = \left[\dfrac{\partial}{\partial x} \quad \dfrac{\partial}{\partial y} \quad \dfrac{\partial}{\partial z} \right]$, $\vec{E}$ is the electric field, $\vec{B}$ is the magnetic field, $\vec{j}$ is the electric current density, which is a vector, and ρ is the electric charge density, which is a scalar (real number).

In empty space, $\vec{j} = \vec{0}$, & $\rho = 0$, and we have:

$$\frac{\partial \vec{E}}{\partial t} = \nabla \times \vec{B}$$

$$\frac{\partial^2 \vec{E}}{\partial t^2} = \frac{\partial}{\partial t}\left(\nabla \times \vec{B}\right) \qquad (19.15)$$

$$= -\nabla \times \frac{\partial \vec{B}}{\partial t}$$

$$= -\nabla \times \left(\nabla \times \vec{E}\right)$$

Using: $curl\left(curl\left(\vec{F}\right)\right) = grad\left(div\left(\vec{F}\right)\right) - \nabla^2 \vec{F}$:

$$\frac{\partial^2 \vec{E}}{\partial t^2} = -\nabla \times \left(\nabla \times \vec{E}\right) \qquad (19.16)$$

$$= -\nabla\left(\nabla \bullet \vec{E}\right) + \nabla^2 \vec{E}$$

But $\nabla \bullet \vec{E} = \rho = 0$

$$\frac{\partial^2 \vec{E}}{\partial t^2} = \nabla^2 \vec{E} \qquad (19.17)$$

This is called the wave equation of the electrical field. With the magnetic field, we have:

$$\frac{\partial \vec{B}}{\partial t} = -\nabla \times \vec{E}$$

$$\frac{\partial^2 \vec{B}}{\partial t^2} = \nabla^2 \vec{B} \qquad (19.18)$$

This is the wave equation for the magnetic field. Putting the c factors into the wave equations leads to electromagnetic waves propagating at the velocity of light, c.

The important thing to notice about these wave equations is that the speed at which the electromagnetic waves propagate is not relative to anything (emitting observer, receiving observer, the sun…).

19.5 MOVING ELECTRIC CHARGES

Working in two dimensions, we apply the space-time rotation matrix (Lorentz transformation) to the electromagnetic matrix. We add in the c to balance the dimensions. Of course, for a stationary charge, the electromagnetic matrix has a zero magnetic part and the Lorentz transformation would be:

$$\begin{bmatrix} c\gamma & v\gamma \\ v\gamma & c\gamma \end{bmatrix} \begin{bmatrix} E & 0 \\ 0 & E \end{bmatrix} = \begin{bmatrix} c\gamma E & v\gamma E \\ v\gamma E & c\gamma E \end{bmatrix} = \gamma \begin{bmatrix} cE & B \\ B & cE \end{bmatrix} \qquad (19.19)$$

We see that a magnetic component has arisen even though there was not one to begin with. This is why magnetic fields exist. The full transformation (rotation) is:

$$\begin{bmatrix} c\gamma & v\gamma \\ v\gamma & c\gamma \end{bmatrix} \begin{bmatrix} E & B \\ B & E \end{bmatrix} = \begin{bmatrix} \gamma cE + v\gamma B & \gamma cB + v\gamma E \\ \gamma cB + v\gamma E & \gamma cE + v\gamma B \end{bmatrix} \qquad (19.20)$$

Note that the above matrix is not within the space-time algebra (the hyperbolic complex numbers) unless $E > B$, but then $c > v$.

Of course, the determinant of the rotation matrix is unity, and so the determinant of the electromagnetic matrix remains unchanged by rotation in space-time. Thus, the real number $E^2 - B^2$ is invariant under change of velocity. We will see this again a few pages later.

In Newtonian mechanics, the gravitational field of a mass, whether moving or not, is spherical. Newtonians would expect the same for the electric field of a point charge. Newtonians would expect that the electrical field surrounding a moving electron would be spherical. To non-Newtonians, the electromagnetic field of a moving charge is subject to length contraction in the direction of motion (and of course to time dilation). Thus, what would be a spherical field emanating from a stationary point charge becomes an ellipsoidal field when the point charge is moving; it becomes a flat disc when the charge moves at the speed of light. A moving bar magnet produces a deformed moving magnetic field and an electric field. A moving point charge produces a deformed moving electric field and a magnetic field. The 3-dimensional equations are:

$$\vec{b} = \frac{1}{c}\vec{v} \times \vec{e}$$

(19.21)

$$\vec{e} = q\frac{\vec{r}}{\gamma r^3 \sin^3 \theta}$$

Where θ is the angle between $\vec{r}$ and the x-axis.

Now let us pause for thought. A material rod is a lot of atoms held in place by electromagnetic forces. If the electromagnetic fields contract in the direction of motion of the rod because of relativistic length contraction, then the length of the rod will contract in the direction of motion. But length contraction is a property of space. Is it possible that the space-time we observe around us is an electromagnetic phenomenon?

19.6   4-TENSORS

This is a book about the theory of special relativity. Special relativity is a 2-dimensional theory. The concepts of that theory can be understood without a detailed mathematical understanding of 4-dimensional electromagnetism. Such a mathematical understanding of electromagnetism necessarily includes 4-tensors. The reader may choose to skip the next few mathematically taxing pages without loss of understanding of special relativity. We include these pages for completeness because electromagnetism is a central part of the universe and special relativity does have a lot to say about it – actually, it is not that mathematically taxing, really.

In the 4-vector presentation of special relativity, we are able to deal with (space and time) dynamics using 4-vectors. The reader might expect that we would be able to deal with electromagnetism using 4-vectors. It is not unreasonable to think that, starting with the three spatial components of the magnetic field, $\vec{B}$, we would be able to add a fourth temporal component to make a 4-vector, $\overline{\overline{B}}$, and similarly with the electric field to form an electric 4-vector, $\overline{\overline{E}}$, but we cannot – we need 4-tensors. We cannot do electromagnetism with two 4-vectors $\{\overline{\overline{E}},\overline{\overline{B}}\}$ because the electric and magnetic fields are entangled together; they are not separate from each other. The type of 4-tensor we need is the "tensor product" of two 4-vectors (which we call a rank two 4-tensor). Such a tensor will have 16 components. (In general, tensors of different types are the tensor products of different numbers of vectors.)

Vectors have invariants, the length of the vector and the angle between two vectors, that do not change under rotation in space (or space-time) – or under any other co-ordinate transformation. Tensors also have invariants that do not change under rotation or under any other co-ordinate transformation. The vectors in space-time are different from the vectors in Euclidean space – they lie over a different type of space - and we calculate the invariants differently (the dot products are different) to suit the different types of space or space-time. The same is true about tensors. The type of tensors that we use in special relativity are called 4-tensors (or Lorentz tensors) analogously to the type of vectors that we use in special relativity being called 4-vectors.

The important thing about tensors is that, if a tensor equation is true in one co-ordinate system, then it is true in all co-ordinate systems (including rotated ones, of course) over that same type of space. Vectors have this property, and it is the case that vectors are a particular type (rank one) of tensor.

19.7 4-POTENTIAL

Although we cannot describe the electromagnetic fields using 4-vectors, we can introduce the concept of a 4-potential into electromagnetism. The reader might recall from an earlier chapter that a potential is something that we differentiate to get the fields. In our case, it is the electromagnetic fields that we want. The reader might also recall that the Aharonov-Bohm experiment done by Chambers indicates that the potential is a real physical thing. Using 3-dimensional vectors, we define a scalar potential (that is a potential with only one component), ϕ, and a vector potential (with three components), $\vec{A}$. We get the fields as:

$$\vec{B} = Curl(\vec{A})$$

$$\vec{E} = -Grad(\phi) - \frac{1}{c}\frac{\partial \vec{A}}{\partial t}$$

(19.22)

Note that, if everything is stationary, then the second of these equations becomes:

$$\vec{E} = -Grad(\phi)$$

(19.23)

This is just the electrostatic field.

Taking the 3-dimensional and scalar potential together, we are able to form a (covariant[9]) 4-vector. Yes! a genuine 4-vector[10].

$$\overline{\overline{O}} = \begin{bmatrix} -\vec{A}_x & -\vec{A}_y & -\vec{A}_z & \varphi \end{bmatrix}$$

(19.24)

The minus signs are what make this 4-vector co-variant. This is basically the same thing as taking the conjugate of a complex number by changing the sign of the imaginary part.

Aside: Where does the electromagnetic 4-potential come from? As well as requiring that physics be unchanged by rotation in space-time, we also require that it be unchanged by rotation in space; this is called $U(1)$ gauge theory. It goes like this:

$U(1)$ invariance is just invariance under rotation in the Euclidean complex plane. We get a $U(1)$ gauge theory by requiring that the Lagrangian be invariant under changes to the field, $\phi(x)$ of the form: (this is just rotation in the complex plane.)

$$\phi(x) \mapsto e^{i\alpha(x)}\phi(x) \quad \& \quad \phi^*(x) \mapsto \phi^*(x)e^{-i\alpha(x)}$$

(19.25)

Where $\alpha(x)$ is a function of space-time. Clearly:

$$\phi^*(x)\phi(x) \mapsto \phi^*(x)e^{-i\alpha(x)}e^{i\alpha(x)}\phi(x) = \phi^*(x)\phi(x)$$

(19.26)

[9.] It does not matter if the reader does not understand the difference between a covariant vector and a contra-variant vector. Your author includes the word for completeness. However, the 4-vectors we met in the dynamics above (4-velocity etc..) are contra-variant vectors whereas the electromagnetic potential is a co-variant 4-vector, as is the Lorentz 4-force.

[10.] Co-variant vectors are often called 1-forms or dual vectors.

However terms involving the derivatives of the field $\phi(x)$ are not automatically invariant under these changes to the field. To make the derivatives $U(1)$ invariant, we are forced to introduce a 4-vector, a_μ. This 4-vector is the electromagnetic potential. And that, we believe is, why electromagnetism exists – because the universe requires invariance under rotation in space!

We are able to do the same sort of thing with quaternion space and the A_3 space that we will meet in later chapters.

Aside: We could add the gradient (of a scalar field) to a vector potential without affecting the curl of the vector potential because the curl of a gradient is zero - $\nabla \times \nabla \phi = 0$. Such an adding of a gradient to a vector potential is called a gauge transformation. Gauge transformations are central to modern theoretical physics, but the idea was first proposed by Hermann Weyl in 1919.[11] That we can add a gradient to a potential without changing the field equations is called gauge invariance; it mitigates the arbitrariness of the potential. Gauge invariance is a property of the Maxwell equations.

By definition, we get the fields as the (sixteen) components of a 4-tensor given by the set of curls:

$$E_{\mu\nu} = \frac{\partial \overline{\overline{O}}_\nu}{\partial \mu} - \frac{\partial \overline{\overline{O}}_\mu}{\partial \nu} \qquad (19.27)$$

Aside: It is common in tensor calculus to indicate differentiation with a comma. (A semi-colon indicates a different type of differentiation.) The above would thus be written:

$$E_{\mu\nu} = \frac{\partial \overline{\overline{O}}_\nu}{\partial \mu} - \frac{\partial \overline{\overline{O}}_\mu}{\partial \nu} = \overline{\overline{O}}_{\nu,\mu} - \overline{\overline{O}}_{\mu,\nu} \qquad (19.28)$$

Wherein μ is either $\{x, y, z, t\}$ and ν is either $\{x, y, z, t\}$; so $\overline{\overline{O}}_x$ is the x component of the 4-potential, for example:

$$E_{tx} = \frac{\partial \overline{\overline{O}}_x}{\partial t} - \frac{\partial \overline{\overline{O}}_t}{\partial x} = \frac{\partial\left(-\vec{A}_x\right)}{\partial t} - \frac{\partial \phi}{\partial t} = -\frac{\partial A_x}{\partial t} - \frac{\partial \phi}{\partial t} \qquad (19.29)$$

[11.] H. Weyl. Ann. Phisik 59, 101 (1919).

The reader might notice that this is just like a curl in 3-dimensional vectors, and we see that the 4-tensor is a set of curls that "fit together properly". There is no "omnipotent mathematical god" that makes the mathematics work out in such a way that the electromagnetic field can be written as a set of curls that "fit together properly"[12]. It happens because space and time "fit together properly" into space-time.

Aside: The electromagnetic field tensor, $E_{\mu\nu}$, can be defined differently from the way it is defined above. The above definition assumes the abelian nature of differentiation. In gauge theory, we do not have this assumption, and the electromagnetic field tensor is defined as:

$$E_{\mu\nu} = \frac{\partial \bar{\bar{O}}_\nu}{\partial \mu} - \frac{\partial \bar{\bar{O}}_\mu}{\partial \nu} - iq\left[\bar{\bar{O}}_\mu, \bar{\bar{O}}_\nu\right] \qquad (19.30)$$

This is referred to as the non-abelian version of the electromagnetic field tensor. This is Yang-Mills theory[13]. The brackets indicate the commutator with which we will deal later. $i = \sqrt{-1}$, and q is electric charge. In electromagnetism, the photon is seen as being an abelian (commutative) potential field. Yang-Mills theory deals with non-abelian potential fields and is a way of extending the electromagnetic mathematical formulation into new areas of physics.

Such is the pedagogic devotion of your author that he writes out the electromagnetic field tensor in full for the edification of his readers. It is normal to forget that tensors formed from two vectors are not strictly speaking a matrix and to pretend that they are a matrix. We will do the same ($E_{row,\,column}$):

[12.] The Maxwell equations for $Div(\bar{B})$ and $Curl(\bar{E})$ are the necessary and sufficient conditions for the existence of a potential.

[13.] Yang Mills theory dates from 1954: C.N.Yang and R.L.Mills Phys. Rev, 96, 191, (1954).

$$E_{\mu\nu} = \begin{vmatrix} \dfrac{\partial \overline{\overline{O}}_1}{\partial x} - \dfrac{\partial \overline{\overline{O}}_1}{\partial x} & \dfrac{\partial \overline{\overline{O}}_2}{\partial x} - \dfrac{\partial \overline{\overline{O}}_1}{\partial y} & \dfrac{\partial \overline{\overline{O}}_3}{\partial x} - \dfrac{\partial \overline{\overline{O}}_1}{\partial z} & \dfrac{\partial \overline{\overline{O}}_4}{\partial x} - \dfrac{\partial \overline{\overline{O}}_1}{\partial t} \\[2mm] \dfrac{\partial \overline{\overline{O}}_1}{\partial y} - \dfrac{\partial \overline{\overline{O}}_2}{\partial x} & \dfrac{\partial \overline{\overline{O}}_2}{\partial y} - \dfrac{\partial \overline{\overline{O}}_2}{\partial y} & \dfrac{\partial \overline{\overline{O}}_3}{\partial y} - \dfrac{\partial \overline{\overline{O}}_2}{\partial z} & \dfrac{\partial \overline{\overline{O}}_4}{\partial y} - \dfrac{\partial \overline{\overline{O}}_2}{\partial t} \\[2mm] \dfrac{\partial \overline{\overline{O}}_1}{\partial z} - \dfrac{\partial \overline{\overline{O}}_3}{\partial x} & \dfrac{\partial \overline{\overline{O}}_2}{\partial z} - \dfrac{\partial \overline{\overline{O}}_3}{\partial y} & \dfrac{\partial \overline{\overline{O}}_3}{\partial z} - \dfrac{\partial \overline{\overline{O}}_3}{\partial z} & \dfrac{\partial \overline{\overline{O}}_4}{\partial z} - \dfrac{\partial \overline{\overline{O}}_3}{\partial t} \\[2mm] \dfrac{\partial \overline{\overline{O}}_1}{\partial t} - \dfrac{\partial \overline{\overline{O}}_4}{\partial x} & \dfrac{\partial \overline{\overline{O}}_2}{\partial t} - \dfrac{\partial \overline{\overline{O}}_4}{\partial y} & \dfrac{\partial \overline{\overline{O}}_3}{\partial t} - \dfrac{\partial \overline{\overline{O}}_4}{\partial z} & \dfrac{\partial \overline{\overline{O}}_4}{\partial t} - \dfrac{\partial \overline{\overline{O}}_4}{\partial t} \end{vmatrix} \quad (19.31)$$

$$= \begin{bmatrix} 0 & \dfrac{\partial \vec{A}_x}{\partial y} - \dfrac{\partial \vec{A}_y}{\partial x} & \dfrac{\partial \vec{A}_x}{\partial z} - \dfrac{\partial \vec{A}_z}{\partial x} & \dfrac{\partial \phi}{\partial x} + \dfrac{\partial \vec{A}_x}{\partial t} \\[2mm] \dfrac{\partial \vec{A}_y}{\partial x} - \dfrac{\partial \vec{A}_x}{\partial y} & 0 & \dfrac{\partial \vec{A}_y}{\partial z} - \dfrac{\partial \vec{A}_z}{\partial y} & \dfrac{\partial \phi}{\partial y} + \dfrac{\partial \vec{A}_y}{\partial t} \\[2mm] \dfrac{\partial \vec{A}_z}{\partial x} - \dfrac{\partial \vec{A}_x}{\partial z} & \dfrac{\partial \vec{A}_z}{\partial y} - \dfrac{\partial \vec{A}_y}{\partial z} & 0 & \dfrac{\partial \phi}{\partial z} + \dfrac{\partial \vec{A}_z}{\partial t} \\[2mm] -\dfrac{\partial \vec{A}_x}{\partial t} - \dfrac{\partial \phi}{\partial x} & -\dfrac{\partial \vec{A}_y}{\partial t} - \dfrac{\partial \phi}{\partial y} & -\dfrac{\partial \vec{A}_z}{\partial t} - \dfrac{\partial \phi}{\partial z} & 0 \end{bmatrix}$$

$$= \begin{bmatrix} 0 & -\vec{B}_z & \vec{B}_y & -\vec{E}_x \\ \vec{B}_z & 0 & -\vec{B}_x & -\vec{E}_y \\ -\vec{B}_y & \vec{B}_x & 0 & -\vec{E}_z \\ \vec{E}_x & \vec{E}_y & \vec{E}_z & 0 \end{bmatrix} \quad (19.32)$$

Aside: If we had chosen the co-ordinate order to be m is either $\{t, x, y, z\}$ and n is either $\{t, x, y, z\}$, we would have come to:

$$F^{\mu\nu} = \begin{bmatrix} 0 & \dfrac{\partial \vec{A}_x}{\partial t} - \dfrac{\partial \phi}{\partial x} & \dfrac{\partial \vec{A}_y}{\partial t} - \dfrac{\partial \phi}{\partial y} & \dfrac{\partial \vec{A}_z}{\partial t} - \dfrac{\partial \phi}{\partial z} \\[2mm] \dfrac{\partial \phi}{\partial x} - \dfrac{\partial \vec{A}_x}{\partial t} & 0 & \dfrac{\partial \vec{A}_y}{\partial x} - \dfrac{\partial \vec{A}_x}{\partial y} & \dfrac{\partial \vec{A}_z}{\partial x} - \dfrac{\partial \vec{A}_x}{\partial z} \\[2mm] \dfrac{\partial \phi}{\partial y} - \dfrac{\partial \vec{A}_y}{\partial t} & \dfrac{\partial \vec{A}_x}{\partial y} - \dfrac{\partial \vec{A}_y}{\partial x} & 0 & \dfrac{\partial \vec{A}_z}{\partial y} - \dfrac{\partial \vec{A}_y}{\partial z} \\[2mm] \dfrac{\partial \phi}{\partial z} - \dfrac{\partial \vec{A}_z}{\partial t} & \dfrac{\partial \vec{A}_x}{\partial z} - \dfrac{\partial \vec{A}_z}{\partial x} & \dfrac{\partial \vec{A}_y}{\partial z} - \dfrac{\partial \vec{A}_z}{\partial y} & 0 \end{bmatrix} \quad (19.33)$$

$$
= \begin{bmatrix}
0 & -\vec{E_x} & -\vec{E_y} & -\vec{E_z} \\
\vec{E_x} & 0 & -\vec{B_z} & \vec{B_y} \\
\vec{E_y} & \vec{B_z} & 0 & -\vec{B_x} \\
\vec{E_z} & -\vec{B_y} & \vec{B_x} & 0
\end{bmatrix}
$$

Which the reader will see in other text books.

The reader should notice that this matrix, $E_{\mu\nu}$, is anti-symmetric, $E_{\mu\nu} = -E_{\nu\mu}$. The 4-tensor of the electromagnetic field is an anti-symmetric tensor. Quaternions are the only wholly anti-symmetric 4-dimensional division algebra. Thus, it ought not to surprise the reader that electromagnetism can be done with quaternions, but, although Maxwell used quaternions when he formulated electromagnetism, he did not do it properly; it is only recently that humankind has figured out how to do it properly.

Note that if the magnetic part, $\vec{A_i}$, of the 4-potential does not vary with time, we get:

$$
\begin{bmatrix}
0 & \dfrac{\partial \vec{A_x}}{\partial y} - \dfrac{\partial \vec{A_y}}{\partial x} & \dfrac{\partial \vec{A_x}}{\partial z} - \dfrac{\partial \vec{A_z}}{\partial x} & \dfrac{\partial \phi}{\partial x} \\[2ex]
\dfrac{\partial \vec{A_y}}{\partial x} - \dfrac{\partial \vec{A_x}}{\partial y} & 0 & \dfrac{\partial \vec{A_y}}{\partial z} - \dfrac{\partial \vec{A_z}}{\partial y} & \dfrac{\partial \phi}{\partial y} \\[2ex]
\dfrac{\partial \vec{A_z}}{\partial x} - \dfrac{\partial \vec{A_x}}{\partial z} & \dfrac{\partial \vec{A_z}}{\partial y} - \dfrac{\partial \vec{A_y}}{\partial z} & 0 & \dfrac{\partial \phi}{\partial z} \\[2ex]
-\dfrac{\partial \phi}{\partial x} & -\dfrac{\partial \phi}{\partial y} & -\dfrac{\partial \phi}{\partial z} & 0
\end{bmatrix}
\tag{19.34}
$$

It is a property of 4-vectors that their three spatial components form a 3-dimensional (spatial) vector. It is a property of anti-symmetric 4-tensors that their six non-zero components will split into two sets of three, each of which is a 3-dimensional vector. The two 3-dimensional vectors are the electric and magnetic field vectors $\{\vec{E}, \vec{B}\}$.

We use notation like:

$$
V_x = \frac{\partial W_z}{\partial y} - \frac{\partial W_y}{\partial z}, \quad V_y = \frac{\partial W_x}{\partial z} - \frac{\partial W_z}{\partial x}, \quad V_z = \frac{\partial W_y}{\partial x} - \frac{\partial W_x}{\partial y}
\tag{19.35}
$$

This allows us to write the electromagnetic field tensor as[14]:

$$
E_{\mu\nu} =
\begin{bmatrix}
0 & -\vec{B}_z & \vec{B}_y & -\vec{E}_x \\
\vec{B}_z & 0 & -\vec{B}_x & -\vec{E}_y \\
-\vec{B}_y & \vec{B}_x & 0 & -\vec{E}_z \\
\vec{E}_x & \vec{E}_y & \vec{E}_z & 0
\end{bmatrix},
$$

$$
E^{\mu\nu} =
\begin{bmatrix}
0 & -\vec{B}_z & \vec{B}_y & \vec{E}_x \\
\vec{B}_z & 0 & -\vec{B}_x & \vec{E}_y \\
-\vec{B}_y & \vec{B}_x & 0 & \vec{E}_z \\
-\vec{E}_x & -\vec{E}_y & -\vec{E}_z & 0
\end{bmatrix}
$$

(19.36)

Aside: The matrix with the subscripted indices is referred to as the covariant electromagnetic 4-tensor. The matrix with the superscripted indices is referred to as the contra-variant electromagnetic 4-tensor. This does not concern us greatly. To change a covariant 4-vector into a contra-variant 4- vector, just alter the sign on the time component.

And so we see that, provided the 4-potential does not vary with time, we can equate the electric field vector with the spatial rate of change of the electrostatic potential, ϕ. The magnetic field vector can be equated with the spatial rate of change of the magnetic vector potential, $\vec{A}$. The reader should realize that we have purely electro-static elements in the tensor only if we treat time as a constant. If we treat time on an equal footing as we treat space, which is inevitable in special relativity, then the "electric field vector" depends on the magnetic vector potential, $\vec{A}$, as well as the electro-static potential, ϕ; in which case, the "electric field vector" is not an "electric field vector" but an electromagnetic field vector. It is only because we stationary observers can ignore time that we have a purely electrical field.

Now, the Lorentz force, $\vec{F} = q\vec{E} + q\vec{v} \times \vec{B}$, depends upon velocity through space – it has the magnetic field multiplied by the velocity. We might expect, and this turns out to be correct, that the electro-magnetic force 4-vector is simply the above field 4-tensor multiplied

---

[14.] This was first done by Minkowski.

by the 4-velocity (and multiplied by the charge, q). This is the 4-vector (we've included a c):

$$4 - Force_{emag} = \frac{q}{c} \begin{bmatrix} 0 & -\vec{B}_z & \vec{B}_y & -\vec{E}_x \\ \vec{B}_z & 0 & -\vec{B}_x & -\vec{E}_y \\ -\vec{B}_y & \vec{B}_x & 0 & -\vec{E}_z \\ \vec{E}_x & \vec{E}_y & \vec{E}_z & 0 \end{bmatrix} \begin{bmatrix} u_x \\ u_y \\ u_z \\ c \end{bmatrix}$$

(19.37)

$$= \frac{q}{c} \begin{bmatrix} -\vec{B}_z u_y + \vec{B}_y u_z - \vec{E}_x c \\ \vec{B}_z u_x - \vec{B}_x u_z - \vec{E}_y c \\ -\vec{B}_y u_x + \vec{B}_x u_y - \vec{E}_z c \\ \vec{E}_x u_x + \vec{E}_y u_y + \vec{E}_z u_z \end{bmatrix}$$

This is, in 4-tensor form, the Lorentz force law. It is normally written as:

$$F_\mu = \frac{q}{c} E_{\mu\nu} U^\nu$$

(19.38)

Aside: The Einstein convention is that

$$E_\nu U^\nu = E_1 U^1 + E_2 U^2 + E_3 U^3 + E_4 U^4 \equiv E \bullet U$$

We can calculate the components of the Lorentz force by multiplying the matrices together, but we'll do the first one more slowly. When $\mu = 1$:

$$F_1 = q E_{1\nu} U^\nu$$
$$= q \left(E_{11} U^1 + E_{12} U^2 + E_{13} U^3 + E_{14} U^4 \right)$$
$$= q \left(0.u_x - \vec{B}_z u_y + \vec{B}_y u_z - \vec{E}_x \right)$$

(19.39)

$$= -q \left(\vec{E}_x + \left(\vec{B}_z u_y - \vec{B}_y u_z \right) \right)$$
$$\vec{F}_x = -q\vec{E} + q\left(\vec{U} \times \vec{B} \right)_x$$

The other bits, by matrix multiplication are:

$$F_2 = -q \left(\vec{E}_y + \left(\vec{B}_x u_z - \vec{B}_z u_x \right) \right)$$
$$F_3 = -q \left(\vec{E}_z + \left(\vec{B}_y u_x - \vec{B}_x u_y \right) \right)$$

(19.40)

Which is the Lorentz force law (with negative signs). We also have:

$$F_4 = -\vec{E} \bullet \vec{U}$$

(19.41)

Which, unfortunately puts the dot product equal to a minus sign (think the norm of the 4-acceleration).

We can put these all together (not concerning ourselves with signs) to form the electromagnetic force (co-variant) 4-vector:

$$\overline{\overline{F}}_{Lorentz} = q\gamma\left[-\vec{F} \quad \frac{1}{c}\vec{E}\bullet\vec{U}\right] = q\gamma\left[-\left(\vec{E}+\frac{1}{c}\vec{U}\times\vec{B}\right) \quad \frac{1}{c}\vec{E}\bullet\vec{U}\right] \quad (19.42)$$

The γ derives from that we differentiate 4-vectors with respect to τ whereas we formed the field tensor by differentiating with respect to the co-ordinates.

We earlier found the 4-acceleration is perpendicular in the space-time sense to the 4-velocity (conventionally but wrongly, their dot product is zero). This implies that the 4-force is perpendicular to the 4-velocity. Let us test this:

$$\begin{bmatrix} -\vec{B}_z u_y + \vec{B}_y u_z - \vec{E}_x c \\ \vec{B}_z u_x - \vec{B}_x u_z - \vec{E}_y c \\ -\vec{B}_y u_x + \vec{B}_x u_y - \vec{E}_z c \\ \vec{E}_x u_x + \vec{E}_y u_y + \vec{E}_z u_z \end{bmatrix} \bullet \begin{bmatrix} u_x \\ u_y \\ u_z \\ c \end{bmatrix} \quad (19.43)$$

$$= u_x\left(-\vec{B}_z u_y + \vec{B}_y u_z - \vec{E}_x c\right) + u_y\left(\vec{B}_z u_x - \vec{B}_x u_z - \vec{E}_y c\right)$$

$$+u_z\left(-\vec{B}_y u_x + \vec{B}_x u_y - \vec{E}_z c\right) + c\left(\vec{E}_x u_x + \vec{E}_y u_y + \vec{E}_z u_z\right) \quad (19.44)$$

$$= u_x\left(-\vec{B}_z u_y + \vec{B}_y u_z\right) + u_y\left(\vec{B}_z u_x - \vec{B}_x u_z\right) + u_z\left(-\vec{B}_y u_x + \vec{B}_x u_y\right)$$

$$= 0$$

We see that we would not have the 4-force perpendicular to the 4-velocity unless we had an anti-symmetric field tensor, or that the "orthogonality" of 4-acceleration and 4-velocity, which derives directly from the nature of space-time, makes the 4-tensor be anti-symmetric. It fits together well!

19.8   THE MAXWELL EQUATIONS II

When we tried to do vector algebra in 4-dimensional space-time using 4-vectors, it worked, but we had to "fudge" a sign in the acceleration 4-vector norm. We begin with the electromagnetic field tensor:

$$E^{\mu\nu} = \begin{bmatrix} 0 & -\vec{B}_z & \vec{B}_y & \vec{E}_x \\ \vec{B}_z & 0 & -\vec{B}_x & \vec{E}_y \\ -\vec{B}_y & \vec{B}_x & 0 & \vec{E}_z \\ -\vec{E}_x & -\vec{E}_y & -\vec{E}_z & 0 \end{bmatrix} \qquad (19.45)$$

We seek $E_{\mu\nu,\nu}$ wherein the comma indicates differentiation with respect to the variable to the right of the comma (ν in this case). This means that we are going to differentiate the νth column with respect to the νth variable (in the order x, y, z, t) from left to right. What we are really doing is differentiating the whole matrix by each variable and then picking out the bits we want from each differentiated matrix. We put the answer equal to the current 4-vector. We get:

$$E^{\mu\nu}{}_{,\nu} = \begin{bmatrix} 0 \\ \dfrac{\partial \vec{B}_z}{\partial x} \\ -\dfrac{\partial \vec{B}_y}{\partial x} \\ -\dfrac{\partial \vec{E}_x}{\partial x} \end{bmatrix} + \begin{bmatrix} -\dfrac{\partial \vec{B}_z}{\partial y} \\ 0 \\ \dfrac{\partial \vec{B}_x}{\partial y} \\ -\dfrac{\partial \vec{E}_y}{\partial y} \end{bmatrix} + \begin{bmatrix} \dfrac{\partial \vec{B}_y}{\partial z} \\ \dfrac{\partial \vec{B}_x}{\partial z} \\ 0 \\ -\dfrac{\partial \vec{E}_z}{\partial z} \end{bmatrix} + \begin{bmatrix} \dfrac{\partial \vec{E}_x}{\partial t} \\ \dfrac{\partial \vec{E}_y}{\partial t} \\ \dfrac{\partial \vec{E}_z}{\partial t} \\ 0 \end{bmatrix} = \begin{bmatrix} -j_x \\ -j_y \\ -j_z \\ -\rho \end{bmatrix} \quad (19.46)$$

We swap all the signs, and we include some constants to suit the units[15].

$$\begin{bmatrix} \dfrac{\partial \vec{B}_z}{\partial y} - \dfrac{\partial \vec{B}_y}{\partial z} \\ \dfrac{\partial \vec{B}_x}{\partial z} - \dfrac{\partial \vec{B}_z}{\partial x} \\ \dfrac{\partial \vec{B}_y}{\partial x} - \dfrac{\partial \vec{B}_x}{\partial y} \\ \dfrac{\partial \vec{E}_x}{\partial x} + \dfrac{\partial \vec{E}_y}{\partial y} + \dfrac{\partial \vec{E}_z}{\partial z} \end{bmatrix} = \mu_0 \begin{bmatrix} j_x + \varepsilon_0 \dfrac{\partial \vec{E}_x}{\partial t} \\ j_y + \varepsilon_0 \dfrac{\partial \vec{E}_y}{\partial t} \\ j_z + \varepsilon_0 \dfrac{\partial \vec{E}_z}{\partial t} \\ \dfrac{\rho}{\varepsilon_0} \end{bmatrix} \qquad (19.47)$$

---

15. $\mu_0 \varepsilon_0 = \dfrac{1}{c^2}$

Which are four of the eight Maxwell equations given above.

Of course, one observer's magnetic field is another observer's electric field. If we swap the signs of the electric field and then swap the electric fields for the magnetic fields, we get the dual electromagnetic tensor. (The reader might want to swap the fields in the Maxwell equations to compare.)

$$B^{\mu\nu} = \begin{bmatrix} 0 & -\vec{E}_z & \vec{E}_y & -\vec{B}_x \\ \vec{E}_z & 0 & -\vec{E}_x & -\vec{B}_y \\ -\vec{E}_y & \vec{E}_x & 0 & -\vec{B}_z \\ \vec{B}_x & \vec{B}_y & \vec{B}_z & 0 \end{bmatrix} \tag{19.48}$$

We do as above to get (which we put equal to zero):

$$B^{\mu\nu}{}_{,\nu} = \begin{bmatrix} \left(\dfrac{\partial \vec{E}_y}{\partial z} - \dfrac{\partial \vec{E}_z}{\partial y} \right) - \dfrac{\partial \vec{B}_x}{\partial t} \\[2mm] \left(\dfrac{\partial \vec{E}_z}{\partial x} - \dfrac{\partial \vec{E}_x}{\partial z} \right) - \dfrac{\partial \vec{B}_y}{\partial t} \\[2mm] \left(\dfrac{\partial \vec{E}_x}{\partial y} - \dfrac{\partial \vec{E}_y}{\partial x} \right) - \dfrac{\partial \vec{B}_z}{\partial t} \\[2mm] \dfrac{\partial \vec{B}_x}{\partial x} + \dfrac{\partial \vec{B}_y}{\partial y} + \dfrac{\partial \vec{B}_z}{\partial z} \end{bmatrix} = \begin{bmatrix} 0 \\ 0 \\ 0 \\ 0 \end{bmatrix} \tag{19.49}$$

Which are the other four of Maxwell's equations. We now have all eight of the Maxwell equations entwined together in the electromagnetic field tensor (and its dual). We can write the Maxwell equations as:

$$E_{\mu\nu,\nu} = \overline{\overline{J}}^{\nu} \qquad\qquad B_{\mu\nu,\nu} = 0 \tag{19.50}$$

Also entwined within the electromagnetic field tensor is the Lorentz force law equation.

If we take the current 4-vector and differentiate each component by the respective variable, as we did for the electromagnetic tensor above, and put it equal to zero, we get the continuity equation:

$$\overline{\overline{J}}^{\nu}{}_{,\nu} = \frac{\partial J_x}{\partial x} + \frac{\partial J_y}{\partial y} + \frac{\partial J_z}{\partial z} + \frac{\partial \rho}{\partial t} = 0 \tag{19.51}$$

This expresses the conservation of electric charge.

19.9 THE COMPONENTS OF ELECTROMAGNETISM

Thus the components of electromagnetism are a 4-velocity, a 4-current, a 4-potential, and a definition of a 4-tensor (and its dual) as a particular set of curls of the 4-potential.

Aside: The Maxwell equations can be written in the Clifford algebra, $cl_3 \equiv Mat(2\mathbb{C})$ or the Clifford algebra $cl_{3,1} \equiv Mat(4\mathbb{R})$. In cl_3, the electric field is taken to be a vector but the magnetic field is taken to be a bi-vector (vectors and bi-vectors are dual in cl_3).

$$\vec{E} = E_1 \vec{e_1} + E_2 \vec{e_2} + E_3 \vec{e_3}$$
$$\vec{B} = B_1 \vec{e_{23}} + B_2 \vec{e_{31}} + B_3 \vec{e_{12}} \tag{19.52}$$

The Maxwell equations are then:

$$\nabla \bullet \vec{E} = \rho$$

$$\frac{\partial \vec{E}}{\partial t} + \nabla \lrcorner \vec{B}\vec{e_{123}} = -\vec{J}$$

$$\frac{\partial \vec{B}}{\partial t} \vec{e_{123}} + \nabla \wedge \vec{E} = 0 \tag{19.53}$$

$$\nabla \wedge \vec{B}\vec{e_{123}} = 0$$

The Maxwell equations have been written as a single equation in many mathematical formulations including as $\mathbb{C}$-vectors[16], as complex quaternions[17], as spinors[18], as Clifford algebras[19], as bi-vectors[20], and as quaternions[21] (see later).

[16.] Silberstein 1907.
[17.] Silberstein 1912–14.
[18.] Laporte & Uhlenbeck 1931.
[19.] Juvet & Schidloof 1932 and Mercier 1935.
[20.] Riesz 1958.
[21.] Peter Jack 2003

Aside: The Yang-Mills non-abelian form of the Maxwell equations is:

$$Div\left(\vec{E}\right) + iq\left(\vec{A} \bullet \vec{E} - \vec{E} \bullet \vec{A}\right) = j_0$$

$$Div\left(\vec{B}\right) + iq\left(\vec{A} \bullet \vec{B} - \vec{B} \bullet \vec{A}\right) = 0$$

$$Curl\left(\vec{E}\right) + \frac{\partial \vec{B}}{\partial t} + iq\left(A_0 B_0 - B_0 A_0 + \vec{A} \times \vec{E} - \vec{E} \times \vec{A}\right) = 0 \quad (19.54)$$

$$\frac{\partial \vec{E}}{\partial t} - Curl\left(\vec{B}\right) + iq\left(A_0 E_0 - E_0 A_0 - \vec{A} \times \vec{B} - \vec{B} \times \vec{A}\right) = \vec{j}$$

We also have the non-abelian current conservation equation:

$$Div\left(\vec{j}\right) - \frac{\partial j_0}{\partial t} - iq\left(A_0 j_0 - j_0 A_0 + \vec{A} \bullet \vec{j} - \vec{j} \bullet \vec{A}\right) = 0 \quad (19.55)$$

In the non-abelian form of electromagnetism, the electric and magnetic fields do not have the property of superposition (that they can be simply added together). In general, the fields from two non-abelian charges cannot be superimposed.

There is also a non-abelian form of the wave equation[22].

EXERCISE

1. How does mass density vary with velocity?

[22.] S. Coleman, Phys. Lett 70B, 59 (1977a).

20

QUATERNIONS

Until now, with the exception of the chapter on electromagnetism, we have been concerned with only 2-dimensional spaces. Although, we looked at 4-vectors, we noted that they are no more than a 2-dimensional mathematical construction with two inert spatial dimensions tagged on for the ride. When we looked at electromagnetism, we were concerned primarily with electromagnetic phenomena and not with the 4-dimensional space of those phenomena. With this chapter, we begin to explore 4-dimensional space-time.

The 4-dimensional space-time in which we sit has several properties which we need to explain:

i. The space in which we sit seems to have a "distance function" of the form: $d^2 = t^2 - x^2 - y^2 - z^2$. It is an algebraic fact that the form of this function is not preserved under multiplication. That is:

$$(e^2 - f^2 - g^2 - h^2)(a^2 - b^2 - c^2 - d^2) \neq T^2 - X^2 - Y^2 - Z^2 \quad (20.1)$$

Therefore, this "distance function" cannot be of the form of the determinant of a rotation matrix. Since distance from the origin is preserved by rotation and the only multiplicative invariant of a matrix is its determinant, the form of the determinant of the rotation matrix of a space is the form of the distance function. Therefore, the above $d^2 = t^2 - x^2 - y^2 - z^2$ cannot be the distance function of a space, or, if it is such a distance function, then it is the distance function of a space with no rotation – which is not really a space. The nature of this purported "distance function" is such that, by setting various pairs of the variables in it to zero, we can reduce the purported 4-dimensional "distance function" to six 2-dimensional distance functions all of which have their form preserved under multiplication and therefore can be proper 2-dimensional geometric spaces. Indeed, we know these six 2-dimensional distance functions to be exactly of the two types of 2-dimensional spaces that derive from the group C_2.

ii. Within electromagnetism, we have that a space-time rotation (velocity) applied to a charged body (an electron, say) produces a magnetic field in a spatial plane perpendicular to the direction of the motion of the charged body. Using a compass to map out this magnetic field shows that it is circular (Euclidean). We have it that a space-time rotation (a velocity boost) produces a spatial rotation; actually, it turns out to be the commutators of the boost that is connected a spatial rotation commutator. We find this phenomenon elsewhere; it is involved in the Thomas precession (see later). It is also intrinsic to the Lorentz group (which we will meet later) that the commutator of a boost (a change of velocity - which is a space-time rotation) multiplied by a boost produces a spatial rotation. The Lorentz group is central to much theoretical physics. These rotation phenomena cannot be 2-dimensional because a 2-dimensional space-time rotation matrix multiplied by a 2-dimensional space-time rotation matrix produces a 2-dimensional space-time rotation matrix and not a purely spatial rotation matrix – the math is simple and irrefutable.

iii. There are clearly 2-dimensional rotations (both types) in the space in which we sit. These two types of 2-dimensional rotations occur in each of three planes. A sensible person might expect only one type of rotation and that it would be a 4-dimensional rotation.

We therefore seem to be sitting in a space that cannot possibly be a space. We have a type of rotation (Thomas precession or electromagnetism) that cannot be 2-dimensional. Nor can it be a 3-dimensional rotation. Although we have not time to go into the details[1], this is because no 3-dimensional Riemannian distance function is closed in form under multiplication. We are left with having to find a 4-dimensional rotation that has the properties we observe. We will find this rotation within the finite group $C_2 \times C_2$. We also need to explain why a function that cannot possibly be the distance function of a space appears to be the distance function of the space in which we sit, and why do we have 2-dimensional rotations. We will also find the answer to this in the group $C_2 \times C_2$. As a bonus, we will also find electro-magnetism, the commutation relations of the Lie group $SU(2)$, and anti-matter in the group $C_2 \times C_2$.

Above, we have written of 4-dimensional rotations. But did not the great mathematician Leonhard Euler (1707–1783) prove that all rotations are 2-dimensional? Yes, he did, but he proved it in spaces of the form $\mathbb{R}^n$. We are not dealing with such spaces. He included the word "all" before rotations because he knew of only spaces of the form $\mathbb{R}^n$ - he had only one eye on what he was doing.

20.1 INTRODUCTION TO QUATERNIONS

We are interested in division algebras to be found in the finite group $C_2 \times C_2$. One such algebra is the well-known quaternions. Since the other algebras in this group are in many ways similar to the quaternions, we will use the familiar quaternions as an example of such algebras.

The quaternions are a 4-dimensional division algebra (4-dimensional type of complex numbers). They have one real axis and three imaginary axes which are three square roots of minus unity (minus-one) called $\{i, j, k\}$. They were discovered by the Irish mathematician William Hamilton (1805–1865) in 1843.

---

[1] The details are to do with the finite group C_3.

Aside*:* In 1844, the exterior product algebras (Grassman algebras) used in particle physics were discovered by Hermann Grassman. In 1845, the octonian algebra (Cayley numbers) was discovered by John Graves and Arthur Cayley[2] (1821–1895). In 1848, James Cockle discovered the hyperbolic complex numbers. It was quite a good five years for algebraists. In 1876, William Kingdon Clifford[3] discovered the Clifford algebras.

James Clerk Maxwell, he of electromagnetism fame, had a great respect for the quaternions, saying, "The invention of the calculus of quaternions by Hamilton is a step towards the knowledge of quantities related to space which can only be compared for its importance with the invention of triple co-ordinates by Descartes[4]."[5]. He wrote on the application of quaternions to electromagnetism in 1870. Maxwell's use of quaternions anticipated the $SU(2)$ weak force and the $SU(2) \times U(1)$ electro-weak unification of particle physics by more than a century, but he used them as 3-dimensional vectors and ignored the real part of the quaternion, thereby rather spoiling it all.

Aside: Clifford algebras could have been combined with Maxwell's quaternion formulation of electromagnetism. This would have led to the Dirac equation of particle mechanics forty years before it was actually written down by Dirac in 1928.

In 1888, Oliver Heaviside and Josiah Willard Gibbs independently rewrote electromagnetism in the present day vector form, and quaternions were largely forgotten until 1962 when Finkelstein, Jaunch, Schiminovich, and Speiser[6] wrote a paper involving quaternions that showed the imaginary quaternion degrees of freedom correspond to the Higgs field that gives mass to the $SU(2)$ gauge bosons. Of course, they did this a year before Higgs invented the Higgs field. The paper also anticipates electro-weak unification well

---

[2] Cayley also formulated matrix algebra.
[3] It was Clifford that translated Riemann's work into English. Riemann's work is the mathematical foundation of general relativity, and so Clifford nearly wrote general relativity 40 years before Einstein.
[4] Descartes (1596–1650).
[5] Volume II of Maxwell's Scientific Papers (pages 570–576).
[6] Helvetica Physica Acta, Vol. XXXV (1962) 328–329.

before Glashow, Salam, and Weinberg. Glashow (1932–), Salam (1926–1996), and Weinberg (1933–) won the 1979 Nobel prize for electro-weak unification.

Unfortunately, modern physicists of the western world take no great interest in quaternions. Most of the published papers are from Mexico or Turkey or China or other non-western countries, and the bi-quaternion formulation of electromagnetism is well known only outside of the western world. An exception to this is Peter Michael Jack who rewrote Maxwell's equations using quaternions (not bi-quaternions) in his paper of 2003[7].

Aside: Within particle physics, we are often concerned with mathematical objects called commutators and anti-commutators. Lie algebra is the study of these things. They are a way of measuring the difference between the non-commutative product of two non-commutative numbers (quaternions say) that arises when the order of the numbers in the product is swapped. The anti-commutator is defined as:

$$\{Q_1, Q_2\} = \frac{1}{2}(Q_1 Q_2 + Q_2 Q_1) \tag{20.2}$$

The commutator is defined as:

$$[Q_1, Q_2] = \frac{1}{2}(Q_1 Q_2 - Q_2 Q_1) \tag{20.3}$$

20.2 QUATERNION MATRIX FORMS

There are two matrix forms of quaternion like algebras. They both derive from the group $C_2 \times C_2$. The first is the anti-quaternions:

$$Q_{Anti} = \begin{bmatrix} a & b & c & d \\ -b & a & d & -c \\ -c & -d & a & b \\ -d & c & -b & a \end{bmatrix} \equiv a + ib + jc + kd \tag{20.4}$$

[7.] Physical space as a quaternion structure, I Maxwell Equations. A Brief Note. arXiv:math-ph/0307038v1 18 Jul 2003.

The b variable is associated with the a square root of minus-one written as i. The c variable is associated with the a second square root of minus-one written as j. The d variable is associated with the a third square root of minus-one written as k. In the Q_{Anti} case, we have the commutation relations:

$$\left\{ \begin{array}{l} i^2 = -1, \; j^2 = -1, \; k^2 = -1, \; ij = -k, \; ji = k, \\ ik = j, \; ki = -j, \; jk = -i, \; kj = i \end{array} \right\} \quad (20.5)$$

Notice the non-commutativity of the $\{i, j, k\}$; we have $ij = -ji$ etc.. Multiplication of the anti-quaternions is just like multiplication of the Euclidean complex numbers, $\mathbb{C}$, but taking account of the commutation relations and being careful to keep everything in order – it is easier to just use the matrix form. These commutation relations are the reverse of the normal quaternion commutation relations, which is why we call them the anti-quaternions or reverse quaternions. The second form is the true quaternions as discovered by William Hamilton:

$$\mathbb{H} = \begin{bmatrix} a & b & c & d \\ -b & a & -d & c \\ -c & d & a & -b \\ -d & -c & b & a \end{bmatrix} \equiv a + ib + jc + kd \quad (20.6)$$

The commutation relations of this algebra are a mirror image of the commutation relations of the anti-quaternion. They are:

$$\left\{ \begin{array}{l} i^2 = -1, \; j^2 = -1, \; k^2 = -1, \; ij = k, \; ji = -k, \\ ik = -j, \; ki = j, \; jk = i, \; kj = -i \end{array} \right\} \quad (20.7)$$

Again, multiplication is simple. Just do it as you would for the $\mathbb{C}$ algebra but take account of the order of everything and of the commutation relations – or use the matrix form. The quaternions and anti-quaternions are the mirror images of each other. They are enantiomorphic spaces (left-handed and right-handed spaces).

Aside: The second of these, the true quaternion, appears in many particle physics books in the form:

$$SU(2) \equiv \begin{bmatrix} a + ib & c + id \\ -c + id & a - ib \end{bmatrix} : a^2 + b^2 + c^2 + d^2 = 1 \quad (20.8)$$

Using the block multiplication properties of matrices and the 2×2 matrix form of the complex numbers, $\mathbb{C}$, will convert this to the 4×4 quaternion matrix above. The Lie group $SU(2)$ is isomorphic to the quaternion rotation matrix and "invariance under $SU(2)$ transformations" can be seen as invariance under rotation in quaternion space. Thus, we can think of the quaternion trigonometric functions as being the trigonometric functions of $SU(2)$ and the quaternion rotation matrix as being $SU(2)$.

Aside: Take the Pauli matrices of particle physics:

$$\sigma_1 = \begin{bmatrix} 1 & 0 \\ 0 & -1 \end{bmatrix}, \quad \sigma_2 = \begin{bmatrix} 0 & -i \\ i & 0 \end{bmatrix}, \quad \sigma_3 = \begin{bmatrix} 0 & 1 \\ 1 & 0 \end{bmatrix} \qquad (20.9)$$

Multiply them each by a real number, add them, and then add a fourth real number matrix to get the quaternion:

$$\begin{bmatrix} a + ib & c + id \\ -c + id & a - ib \end{bmatrix} \qquad (20.10)$$

The (anti-symmetric) rotation matrix of the quaternions is obtained by exponentiating the quaternion matrix (with zero real part). You are about to see a truly 4-dimensional rotation:

$$Q_{Rot} = \begin{bmatrix} \cos(\lambda) & \frac{b}{\lambda}\sin(\lambda) & \frac{c}{\lambda}\sin(\lambda) & \frac{d}{\lambda}\sin(\lambda) \\ -\frac{b}{\lambda}\sin(\lambda) & \cos(\lambda) & -\frac{d}{\lambda}\sin(\lambda) & \frac{c}{\lambda}\sin(\lambda) \\ -\frac{c}{\lambda}\sin(\lambda) & \frac{d}{\lambda}\sin(\lambda) & \cos(\lambda) & -\frac{b}{\lambda}\sin(\lambda) \\ -\frac{d}{\lambda}\sin(\lambda) & -\frac{c}{\lambda}\sin(\lambda) & \frac{b}{\lambda}\sin(\lambda) & \cos(\lambda) \end{bmatrix} \quad (20.11)$$

$$\lambda = \sqrt{b^2 + c^2 + d^2}$$

Within this rotation matrix are the quaternion trigonometric functions[8]. They are similar to the 2-dimensional Euclidean trigonometric functions because all the "imaginary" variables are anti-symmetric

---

[8.] It is not really within the remit of this book, but I should mention that the square root sign leads to "twice as much rotation" as we have in the complex plane. See: Dennis Morris The Physics of Empty Space ISBN: 978-1507707005.

and because $C_2 \times C_2$ has three order two sub-groups (one for each "imaginary" variable). This is quaternion space, and this rotation matrix is the 4-dimensional rotation matrix in that 4-dimensional quaternion space. It is something that you have probably never imagined and probably did not know existed. Try setting some of the variables to zero and playing with commutators of the products of the quaternion rotation matrix.

Every geometric space has within it the concept of angle and of rotation. Is it such a surprise that a 4-dimensional space should have 4-dimensional rotations? As pointed out above, we need them to understand observed physical phenomena. Rotation in quaternion space (also known as $SU(2)$) is utterly central to quantum field theory – it is all over the place in particle physics.

Aside: The unit quaternions are the Lie group $SU(2)$. This group is denoted in many different ways. We have: $Q_{Rot} = Sp(1) = SU(2) = Spin(3) = S3$. There is, however, a difference between Q_{Rot} and $SU(2)$. $SU(2)$ is conceived as existing in $\mathbb{C}^2$ (two copies of the complex plane fitted together) space whereas Q_{Rot} exists in quaternion space.

The quaternion rotation matrix is non-commutative.[9] Taking the commutator of individual variables gives the Lie algebra. Having met 4-dimensional rotation in a 4-dimensional finite group space that derived from the group $C_2 \times C_2$, we are now ready to look at the other 4-dimensional spaces in this group.

20.3   THE $C_2 \times C_2$ FINITE GROUP AND THE SPACE IN WHICH WE SIT

We earlier worked with the algebras (geometric spaces) that derive from the finite group C_2. Of course, these spaces are 2-dimensional but, it seems, we live in a 4-dimensional universe. We are now to work with the geometric spaces that derive from the order four

---

[9.] Rotation on the left takes you to a different place than rotation on the right. One angle but two different rotations.

group $C_2 \times C_2$. This, as it appears, is a group formed by crossing C_2 with itself. Thus, in a sense, we are still in the C_2 universe. We will not be looking at the other order four finite group, C_4.

We can look at the group C_2 as being the multiplicative relations between the numbers $\{(+1), (-1)\}$. We can look at the group $C_2 \times C_2$ as being the multiplicative relations between the pairs of numbers:

$$\left\{ \begin{pmatrix} +1 \\ +1 \end{pmatrix}, \begin{pmatrix} -1 \\ -1 \end{pmatrix}, \begin{pmatrix} -1 \\ +1 \end{pmatrix}, \begin{pmatrix} +1 \\ -1 \end{pmatrix} \right\} \tag{20.12}$$

with multiplication component-wise.

Aside: $C_2 \times C_2 \times C_2$ is the multiplicative relations between:

$$\left\{ \begin{pmatrix} +1 \\ +1 \\ +1 \end{pmatrix}, \begin{pmatrix} -1 \\ +1 \\ +1 \end{pmatrix}, \begin{pmatrix} +1 \\ -1 \\ +1 \end{pmatrix}, \begin{pmatrix} +1 \\ +1 \\ -1 \end{pmatrix}, \begin{pmatrix} -1 \\ -1 \\ +1 \end{pmatrix}, \begin{pmatrix} +1 \\ -1 \\ -1 \end{pmatrix}, \begin{pmatrix} -1 \\ +1 \\ -1 \end{pmatrix}, \begin{pmatrix} -1 \\ -1 \\ -1 \end{pmatrix} \right\} \tag{20.13}$$

The matrix form of the $C_2 \times C_2$ group is:

$$\begin{bmatrix} a & b & c & d \\ b & a & d & c \\ c & d & a & b \\ d & c & b & a \end{bmatrix} \tag{20.14}$$

There are sixteen division algebras that derive from $C_2 \times C_2$. Eight (two of type A_1 and six of type A_2) of these division algebras are commutative division algebras (algebraic fields).

The A_1 algebra contains three square roots of plus unity (plus-one). This compares to the hyperbolic complex numbers, $\mathbb{S}$, that have one square root of plus unity and the Euclidean complex numbers, $\mathbb{C}$, that have one square root of minus unity (minus-one). The A_2 algebra contains one square root of plus unity and two square roots of minus unity. We have no interest in these commutative algebras. It seems that the 4-dimensional universe, or at least the observable part of it, is comprised of only non-commutative algebras[10].

[10.] What we did with C_2 can be equally well done with $C_2 \times C_2$ by setting two of the imaginary variables to zero. It was pedagogically easier to use C_2.

Aside: The Clifford algebra, cl_2, has three square roots of plus unity and one square root of minus unity. The Clifford algebra, cl_3, has four square roots of plus unity and four square roots of minus unity. The Clifford algebra, cl_4, has six square roots of plus unity and ten square roots of minus unity.

20.4 NON-COMMUTATIVE SPACES

In the two dimensional complex plane, a rotation through θ degrees followed by a rotation through ϕ degrees gets to the same point as a rotation through ϕ degrees followed by a rotation through θ degrees. This is because the rotation matrices of these two rotations are commutative; it does not matter in which order the product of the matrices is written:

$$\begin{bmatrix} \cos\theta & \sin\theta \\ -\sin\theta & \cos\theta \end{bmatrix}\begin{bmatrix} \cos\phi & \sin\phi \\ -\sin\phi & \cos\phi \end{bmatrix} = \begin{bmatrix} \cos\phi & \sin\phi \\ -\sin\phi & \cos\phi \end{bmatrix}\begin{bmatrix} \cos\theta & \sin\theta \\ -\sin\theta & \cos\theta \end{bmatrix} \quad (20.15)$$

Now imagine a 3-dimensional sphere. It has three planes of rotation (the slice parallel to the Greenwich meridian, the slice parallel to longitude 90° west and the slice parallel to the equator). With a little thought, the reader will see that rotation in one of these planes followed by rotation in a different plane does not necessarily get the same result as when the rotations are done in the reverse order. This is because the rotation matrices are not commutative:

$$\begin{bmatrix} \cos\theta & \sin\theta & 0 \\ -\sin\theta & \cos\theta & 0 \\ 0 & 0 & 1 \end{bmatrix}\begin{bmatrix} 1 & 0 & 0 \\ 0 & \cos\phi & \sin\phi \\ 0 & -\sin\phi & \cos\phi \end{bmatrix}$$
$$\neq \begin{bmatrix} 1 & 0 & 0 \\ 0 & \cos\phi & \sin\phi \\ 0 & -\sin\phi & \cos\phi \end{bmatrix}\begin{bmatrix} \cos\theta & \sin\theta & 0 \\ -\sin\theta & \cos\theta & 0 \\ 0 & 0 & 1 \end{bmatrix} \quad (20.16)$$

We say that the 2-dimensional space, $\mathbb{C}$, is commutative and that the 3-dimensional space, $\mathbb{R}^3$, is non-commutative. Of course, as shown by the three 3×3 "rotation matrices", the sphere does not exist in 3-dimensional space but in three 2-dimensional spaces.

The other eight division algebras that derive from the $C_2 \times C_2$ are non-commutative. That is worth emphasizing:

The commutative group $C_2 \times C_2$ contains eight non-commutative division algebras.

We give two examples:

$$\mathbb{H} = \begin{bmatrix} a & b & c & d \\ -b & a & -d & c \\ -c & d & a & -b \\ -d & -c & b & a \end{bmatrix} \qquad A_3 = \exp\left(\begin{bmatrix} a & b & c & d \\ -b & a & -d & c \\ c & -d & a & -b \\ d & c & b & a \end{bmatrix}\right) \qquad (20.17)$$

There are two types of quaternions (the quaternions and the anti-quaternions) and six types of A_3 division algebras. Like the quaternions, the A_3 algebras have commutation relations. These commutation relations are different in each algebra and different from the quaternions and the anti-quaternions. The A_3 algebras come in three pairs with each pair being comprised of an algebra and its anti-algebra. The commutation relations of an anti-algebra are the reverse of the commutation relations of the algebra. The distribution of the minus signs in the bottom right 3×3 corner of the 4×4 matrix are reversed in the anti-algebra.

It is quite amazing that the commutative group $C_2 \times C_2$ contains non-commutative division algebras. (The reason for this is that real numbers have both positive and negative square roots, and although your author would like to show the reader why non-commutativity exists, it is too much of a digression from the subject of this book.)

A_3 contains one square root of minus unity and two square roots of plus unity. The quaternions, of course, contain three square roots of minus unity. What balance we have within these four types of $C_2 \times C_2$ spaces; how fairly they share the square roots of plus and minus unity between them. This reflects, and is the same thing as, how fairly they share the different types of 2-dimensional rotations (Euclidean spatial or space-time boosts) between them. The mixed symmetric and anti-symmtric A_3 rotation matrix, of the form (again an example of one of the six):

$$A_{3Rot} = \begin{bmatrix} \cosh(\lambda) & \frac{b}{\lambda}\sinh(\lambda) & \frac{c}{\lambda}\sinh(\lambda) & \frac{d}{\lambda}\sinh(\lambda) \\ -\frac{b}{\lambda}\sinh(\lambda) & \cosh(\lambda) & -\frac{d}{\lambda}\sinh(\lambda) & \frac{c}{\lambda}\sinh(\lambda) \\ \frac{c}{\lambda}\sinh(\lambda) & -\frac{d}{\lambda}\sinh(\lambda) & \cosh(\lambda) & -\frac{b}{\lambda}\sinh(\lambda) \\ \frac{d}{\lambda}\sinh(\lambda) & \frac{c}{\lambda}\sinh(\lambda) & \frac{b}{\lambda}\sinh(\lambda) & \cosh(\lambda) \end{bmatrix} \quad (20.18)$$

$$\lambda = \sqrt{-b^2 + c^2 + d^2}$$

Notice that the b variable is anti-symmetric within this matrix whereas the c and d variables are symmetric. This means both types of 2-dimensional rotation are within this rotation matrix. This means both types of 2-dimensional rotation are in this space! The other A_3 rotation matrices are basically the same but with a different anti-symmetric variable.

The A_3 rotation matrix is a 4-dimensional rotation matrix – it does 4-dimensional rotation – just like the quaternion rotation matrix. None-the-less, I think emphasis is warranted:

4 × 4 rotation matrices do 4-dimensional rotation.

The important thing about the A_3 algebras is that they contain a mixture of both symmetric and anti-symmetric "imaginary" variables. They each have two symmetric "imaginary" variables and one anti-symmetric "imaginary" variable. This means that A_3 space (all six types) contains both symmetric rotations (rotations in space-time, also called boosts) and an anti-symmetric rotation (spatial rotation) – remember the phrase from the Lorentz group "the commutator of a boost times a boost makes a spatial rotation". We need both types of 2-dimensional rotation in one space.

The $C_2 \times C_2$ group has three order two subgroups each formed from the real variable and one of the imaginary variables. There are thus three 2-dimensional rotations within the A_3 space. By simple multiplication, we see that the commutator of the product of the c variable and the d variable in the above A_3 matrix is the b variable, and we have that the commutator of two space-time rotations (symmetric rotations) multiplied together is a spatial rotation (anti-symmetric rotation).

The b variable corresponds to a Euclidean rotation, when $c = d = 0$, and the $\{c, d\}$ variables correspond to space-time rotations when either of them is zero and $b = 0$. This is just as in the Lorentz group. In the A_3 algebra above, we have that the commutator of the symmetric (space-time) $\{c, d\}$ variables is the anti-symmetric (Euclidean space) b variable. The six isomorphic forms of the A_3 algebra permute the axes, but they all have that the commutator of the product of two space-time rotations is a Euclidean space rotation.

20.5 THE SPACE IN WHICH WE SIT

We are about to derive the 4-dimensional space-time in which we sit. It brings with it the field equations of general relativity. We will skimp on the general relativity bit because general relativity is outside of the remit of this book.

The six A_3 algebras come in three pairs; each pair has one A_3 algebra and one A_3 anti-algebra. We have six copies of the same algebra written in six different bases. There is a distance function in each of the six A_3 algebras. We take the expectation distance function (like expectation values in quantum mechanics) by adding the six distance functions:

$$\text{Sum} \begin{pmatrix} t^2 + x^2 - y^2 - z^2 \\ t^2 + x^2 - y^2 - z^2 \\ t^2 - x^2 + y^2 - z^2 \\ t^2 - x^2 + y^2 - z^2 \\ t^2 - x^2 - y^2 + z^2 \\ t^2 - x^2 - y^2 + z^2 \end{pmatrix} = 6t^2 - 2x^2 - 2y^2 - 2z^2 \qquad (20.19)$$

The ratio of 6 to 2 is just the units in which we measure time and space. We have the distance function of our space-time.

We take the expectation algebra by adding the six algebras, but wait, you cannot add isomorphic algebras written in different bases without destroying the algebras. If we destroy the algebras, we have no multiplication operation[11]. With no multiplication operation, we have no rotation (that's why we do not see 4-dimensional rotation is our space-time) and we have no imaginary variables; all that remains is four, not imaginary, variables; they must be real variables. These four real variables are a 4-dimensional manifold; we have $\mathbb{R}^4$. A classical type of space has emerged from the finite group spaces as an expectation space.

We have a 4-dimensional manifold with a distance function. Insisting that a vector maintains its length at different points in the manifold (as measured by the distance function) induces a metric tensor on to the manifold. Allowing a A_3 phase (angle) to vary from point to point in the manifold induces an affine connection (a notion of parallel transport) on to the manifold. (With the distance function, this is the Levi-Civita connection.)

We have our 4-dimensional space-time. The second differentials of the metric tensor form a unique tensor called the Riemann curvature tensor. From this we get the Einstein tensor. A second rank symmetric tensor naturally emerges from the sum of the symmetric parts of the fields of the A_3 algebras. We assume this is the mass-energy tensor and put it equal to the Einstein tensor[12] – we have the field equations of general relativity.

When, by observation, we take the expectation algebra, we smash the A_3 algebras and we smash the 4-dimensional rotation. There is no mathematical connection between the A_3 algebras and the emerging $\mathbb{R}^4$ space. Thus there is no deterministic connection between these two parts of the universe. It seems that the A_3 algebras are quantum gravity. We presume that they each have a charge of mass. Since there are three pairs of them, we would expect particles to have three different masses. We call this three generations of particles.

Because the mathematics produces only one copy of each of the 2-dimensional rotations, taking the 2-dimensional expectation algebras leaves the algebras unchanged; we keep 2-dimensional rotations. Because the 2-dimensional distance functions are sub-dis-

---

[11.] "Proper" multiplication exists in only division algebras.

[12.] We have guessed the field equations, but then so did Einstein.

tance functions of our 4-dimensional space-time distance function, we have the two types of 2-dimensional rotation within our 4-dimensional space-time.

I short, we postulate that the space in which we sit is the sum of all the six A_3 algebras and both 2-dimensional algebras.

Thus it is that, if we add the A_3 spaces, we get the distance function of the space in which we sit. We also have the 4-dimensional rotations that we need to explain electromagnetic phenomena, the Lorentz group we are yet to meet, and the Thomas precession, and other, very technical, stuff, and we have the 2-dimensional rotations that we observe. We also have 2-dimensional rotations in three 2-dimensional planes within these spaces. So, it seems, we live in the sum of six $C_2 \times C_2$ finite group spaces[13]. Of course, these spaces exist in only their polar forms, but that too is what we observe, there is a limiting velocity (all equal) in all three spatial directions. Job done!

And all from nothing more than the real numbers and the multiplicative relations between $\{+1, -1\}$; oh! and the exponential function, of course. It gives good reason to pause for thought.

EXERCISE

1. Within both the quaternion algebra and the anti-quaternion algebra (two separate calculations), calculate the product: $(a + ib + jc + kd)(t + ix + jy + kz)$. Remember the terms in order and see. If you reverse the order of the terms in the product, do you get the same answers?

[13.] These are spinor spaces.

21

ELECTROMAGNETISM WITH QUATERNIONS

This book is about space-time. In the previous chapter, we found the 4-dimensional space-time in which we sit to be formed from the sum of the six A_3 algebras. These algebras are derived from the group $C_2 \times C_2$. Also derived from that group are the quaternions and the anti-quaternions; what role do they play in the universe? The quaternions and anti-quaternions give rise to the Maxwell equations of electromagnetism and to the anti-Maxwell equations of anti-electromagnetism (anti-matter). We give a, rather brief, overview of that here.[1]

There is an almost religious belief among physicists that the universe will use "simple" and "elegant" mathematics. We expect that, when we eventually discover the "Grand Unified Field Theory", it will be beautiful, elegant, and simple. If we can build a theory from no more than only numbers (of various types) derived from the finite groups, then that theory will "have" to be correct because numbers are correct. When we work with $\mathbb{R}^n$, we are doing "inelegant" math. It is humankind, by the act of taking expectation spaces through observation, that fits copies of the finite group spaces together to form $\mathbb{R}^n$; the $\mathbb{R}^n$ construction is not found in mathematics but is a consequence of observation by humankind – that is what is "inelegant" about it. The quaternions do exist in mathematics. They are not

---

[1] This is thoroughly covered in Dennis Morris : The Physics of Empty Space ISBN: 978-1507707005

invented by humankind but derive from the $C \times C_2$ finite group. Thus, the idea is that we can derive the whole of physics simply by doing the mathematics properly (if we knew how). In the previous chapter, we got the space in which we sit from no more than numbers. Below, we get classical electromagnetism together with the $SU(2)$ commutation relations and anti-matter, from nothing more than numbers, and so we are now doing number theory but calling it electromagnetism.

We remind the reader that the commutator and anti-commutator of two quaternions are defied as:

$$\{Q_1, Q_2\} = \frac{1}{2}(Q_1 Q_2 + Q_2 Q_1) \tag{21.1}$$

When differentiating quaternion vector fields, we need to take account of their non-commutative nature. We have left-differentiation (pre-multiplied), d_L, and we have right-differentiation (post-multiplied), d_R.

For the differential of a quaternion vector potential, Φ_Q, we have:

$$\{\Phi_Q, d\} = \frac{1}{2}\left(d_{Right}\left(\Phi_Q\right) + d_{Left}\left(\Phi_Q\right)\right)$$
$$[\Phi_Q, d] = \frac{1}{2}\left(d_{Right}\left(\Phi_Q\right) - d_{Left}\left(\Phi_Q\right)\right) \tag{21.2}$$

The choice of subtracting the left differential rather than the right differential is arbitrary. The reader will find the opposite convention elsewhere.

We differentiate quaternion electromagnetic potential matrix with respect to a quaternion matrix. We do not know where this potential field comes from (unless it is quaternion phase varying locally over space-time like the phase of the wave-function in $U(1)$ gauge theory). We just assume it exists. We take the differentials to get:

The electric field vector:

$$\{\Phi_Q, d\} = \begin{bmatrix} T & \vec{E}_x & \vec{E}_y & \vec{E}_z \\ -\vec{E}_x & T & -\vec{E}_z & \vec{E}_y \\ -\vec{E}_y & \vec{E}_z & T & -\vec{E}_x \\ -\vec{E}_z & -\vec{E}_y & \vec{E}_x & T \end{bmatrix} = E_Q \tag{21.3}$$

$$T = \frac{\partial \phi}{\partial t} + \frac{\partial A_x}{\partial x} + \frac{\partial A_y}{\partial y} + \frac{\partial A_z}{\partial z}$$

And the magnetic field vector:

$$
\left[\Phi_Q, d\right] =
\begin{bmatrix}
0 & \vec{B}_x & \vec{B}_y & \vec{B}_z \\
-\vec{B}_x & 0 & -\vec{B}_z & \vec{B}_y \\
-\vec{B}_y & \vec{B}_z & 0 & -\vec{B}_x \\
-\vec{B}_z & -\vec{B}_y & \vec{B}_x & 0
\end{bmatrix}
= B_Q
\tag{21.4}
$$

Thus, the potential splits into two separate vector fields because of the non-commutativity of the differentiation. The two separate vector fields are the electric field and the magnetic field. Let us just pause to consider that. The quaternions are telling us that the reason we have two types of vector field in electromagnetism, the electric vector and the magnetic vector, is because the quaternions are non-commutative and hence have non-commutative differentiation. That is probably worth the pause that we took.

Of course, in the conventional 4-vector presentation of electromagnetism, we were driven to combine two vectors, the electric field and the magnetic field, into one 4-tensor. Here, in the quaternions, we have those two vectors separate but tied together.

$$
\left[E_Q, d\right] = \left\{B_Q, d\right\}
\tag{21.5}
$$

This gives:

$$
\begin{bmatrix}
0 & \sim & \sim & \sim \\
\dfrac{\partial \vec{E}_y}{\partial z} - \dfrac{\partial \vec{E}_z}{\partial y} & \sim & \sim & \sim \\
\dfrac{\partial \vec{E}_z}{\partial x} - \dfrac{\partial \vec{E}_x}{\partial z} & \sim & \sim & \sim \\
\dfrac{\partial \vec{E}_x}{\partial y} - \dfrac{\partial \vec{E}_y}{\partial x} & \sim & \sim & \sim
\end{bmatrix}
=
\begin{bmatrix}
-\dfrac{\partial B_x}{\partial x} - \dfrac{\partial B_y}{\partial y} - \dfrac{\partial B_z}{\partial z} & \sim & \sim & \sim \\
\dfrac{\partial B_x}{\partial t} & \sim & \sim & \sim \\
\dfrac{\partial B_y}{\partial t} & \sim & \sim & \sim \\
\dfrac{\partial B_z}{\partial t} & \sim & \sim & \sim
\end{bmatrix}
\tag{21.6}
$$

These are the four homogeneous Maxwell equations. Thus, using quaternions, we have four of the "Maxwell equations" of classical electromagnetism as:

$$
\left\{B_Q, d\right\} = \left[E_Q, d\right]
\tag{21.7}
$$

Doing the same with the anti-quaternions produces another set of four homogeneous Maxwell equations.

The homogeneous Maxwell equations are no more than the differential identities of quaternion and anti-quaternion differentiation.

21.1 WHY NO ANTI-MATTER IN OUR CLASSICAL UNIVERSE

Taking the expectation fields (that is adding the four fields together – the quaternion electric field, the quaternion magnetic field, the anti-quaternion electric field, the anti-quaternion magnetic field) leads to a quaternion distribution of minus signs in the electromagnetic tensor. We do not get the anti-quaternion distribution of minus signs in the electromagnetic tensor. We therefore live in a matter universe rather than an anti-matter universe. We find anti-matter in only the finite group spaces.

21.2 THE INHOMOGENEOUS MAXWELL EQUATIONS

When we take the expectation Maxwell equations by adding the Maxwell equations of both the quaternions and the anti-quaternions, we find that, in order to fit these into the quaternion form of the electromagnetic tensor, we have to put parts of them equal. This gives us the four inhomogeneous Maxwell equations.

The homogeneous Maxwell equations are just a differential identity and are true in classical space and in the finite group spaces. The inhomogeneous Maxwell equations are true in only classical space.

The inhomogeneous Maxwell equations are:

$$
\begin{bmatrix}
0 & \sim \\
\dfrac{\partial B_z}{\partial y} - \dfrac{\partial B_y}{\partial z} & \sim \\
-\dfrac{\partial B_x}{\partial z} - \dfrac{\partial B_z}{\partial x} & \sim \\
\dfrac{\partial B_y}{\partial x} - \dfrac{\partial B_x}{\partial y} & \sim
\end{bmatrix}
=
\begin{bmatrix}
-\left(\dfrac{\partial T}{\partial t} + \dfrac{\partial \vec{E}_x}{\partial x} + \dfrac{\partial \vec{E}_y}{\partial y} + \dfrac{\partial \vec{E}_z}{\partial z} \right) & \sim \\
\dfrac{\partial \vec{E}_x}{\partial t} - \dfrac{\partial T}{\partial x} & \sim \\
\dfrac{\partial \vec{E}_y}{\partial t} - \dfrac{\partial T}{\partial y} & \sim \\
\dfrac{\partial \vec{E}_z}{\partial t} - \dfrac{\partial T}{\partial z} & \sim
\end{bmatrix}
\tag{21.8}
$$

T is the electric field in the time direction. If space and time are not different things but just two dimensions in space-time, we would expect vector fields to have components in the time direction; of course, the magnetic field vector has a zero time component. We put:

$$j_x = -\frac{\partial T}{\partial x}, \; j_y = -\frac{\partial T}{\partial y}, \; j_z = -\frac{\partial T}{\partial z}, \; \frac{\rho}{\varepsilon_0} = -\frac{\partial T}{\partial t} \qquad (21.9)$$

This seems to invite the interpretation that the electric charge density, ρ, is the electric current in time direction – everything moves forward in time. Throwing in some constants gives:

$$\frac{\partial \vec{E}_x}{\partial x} + \frac{\partial \vec{E}_y}{\partial y} + \frac{\partial \vec{E}_z}{\partial z} = \frac{\rho}{\varepsilon_0}$$

$$-\begin{bmatrix} \dfrac{\partial B_z}{\partial y} - \dfrac{\partial B_y}{\partial z} \\[2mm] \dfrac{\partial B_x}{\partial z} - \dfrac{\partial B_z}{\partial x} \\[2mm] \dfrac{\partial B_y}{\partial x} - \dfrac{\partial B_x}{\partial y} \end{bmatrix} = \mu_0 \begin{bmatrix} \varepsilon_0 \dfrac{\partial \vec{E}_x}{\partial t} + j_x \\[2mm] \varepsilon_0 \dfrac{\partial \vec{E}_y}{\partial t} + j_y \\[2mm] \varepsilon_0 \dfrac{\partial \vec{E}_z}{\partial t} + j_z \end{bmatrix} \qquad (21.10)$$

Aside: Of course, the quaternions and the anti-quaternions also have the $SU(2)$ commutation relations of quantum field theory. Thus, we have done as well as Maxwell plus anti-matter plus quantization.

Maxwell's equations can be written in several different mathematical constructions including 4-vector formulation and Clifford algebra formulation.

CHAPTER 22

4-DIMENSIONAL SPACE-TIME AND THE LORENTZ GROUP

In the earlier parts of this book, we dealt with the theory of special relativity in only 2-dimensional space. This was pedagogically convenient, and, since special relativity is essentially a 2-dimensional theory, we lost nothing of the theory by so doing. Doing that, we came to understand time dilation, mass increase, the unity of momentum and energy etc... However, we clearly live in a space that is 4-dimensional. This book is primarily about the nature of empty space and time and motion in that empty space. We are driven to the theory of special relativity because it describes so much of that nature, but the theory of special relativity begins by assuming *a priori* the existence, substantiated by observation, of 4-dimensional space-time. In this book, along with other things, we are trying to understand why space has the nature that we observe. Why it is 4-dimensional? and why are three of those dimensions spatial? and why does it exist?

We have seen that the finite group $C_2 \times C_2$ has within it the quaternion division algebras and the A_3 division algebras. We have seen that the quaternions have within them the Maxwell equations of electromagnetism. We have seen that by adding the six A_3 spaces on to the same axes, we get something that seems to match the 4-dimensional space-time in which we sit. We note that these spaces each contain 4-dimensional rotations. In this chapter, we present the conventionally accepted description of 4-dimensional space-time known as the Lorentz group, and we see that it does contain 4-dimensional rotations and that it matches the sum of the A_3 spaces.

22.1   THE LORENTZ GROUP

The Lorentz group is a representation of the 4-dimensional space-time in which we seem to live. It is conventionally based in $\mathbb{R}^4$. Many texts define the Lorentz group as the set of linear transformations that leave the space-time interval $s^2 = c^2t^2 - x^2 - y^2 - z^2$ invariant. Any linear transformation that leaves a distance function invariant is a rotation, and we might expect that the Lorentz group would be a 4×4 rotation matrix, but this is impossible because the form of the quadratic distance function with signature $(+ -, -, -)$ is not preserved under multiplication. This tells us that 4-dimensional space-time is not a single space. Nor is it possible to consider the Lorentz group to be a set of 3×3 rotation matrices, for the same reason as above. The Lorentz group is therefore based upon the set of 2-dimensional rotations in the 4-dimensional space-time that we seem to inhabit. We have seen that there are only two types of 2-dimensional rotation. In the 4-dimensional space-time we observe, there are six 2-dimensional planes and six corresponding 2-dimensional rotations; that is three temporal rotations in the $\{(t, x), (t, y), (t, z)\}$ planes, and three spatial rotations in the $\{(x, y), (x, z), (y, z)\}$ planes. We believe, and this is verified by observation, that the physics of the universe is invariant under these rotations. Indeed, it is an axiom of field theory that every element of the Lorentz group (which we call $SO(3, 1)$) is a symmetry of the field theory. This is worth emboldening.

22.2   PHYSICS IS INVARIANT UNDER ALL ROTATIONS IN THE LORENTZ GROUP

The Lorentz group associates each of the six 2-dimensional rotations with a 4×4 matrix, but it does so by throwing lots of zeros into what would otherwise be a 2×2 matrix. The Lorentz group is expressed as a set of commutation relations between these six matrices. It is a fact that, within the non-commutative 4-dimensional finite group spaces, we have to use commutation relations to express relations between 2-dimensional rotations in the 2-dimensional sub-spaces of those 4-dimensional finite group spaces. In this, the 4-dimensional finite group spaces are already looking like the Lorentz group.

Aside: The finite group C_3 holds geometric spaces with 3-dimensional rotations. The finite group C_4 holds geometric spaces with both 4-dimensional rotations and a 2-dimensional rotation. The finite group C_5 holds geometric spaces with 5-dimensional rotations, etc... Sets of more than one 2-dimensional rotation occur in only the $C_4 \times C_4 \times \ldots$ groups and in the $\mathbb{R}^n$, $\mathbb{C}^n$, $\mathbb{H}^n$ spaces.

Since the 2-dimensional rotation matrices derive from the exponentiation of a variable, b, which is symmetric in the hyperbolic case and anti-symmetric in the Euclidean case, we refer to the two types of rotation as symmetric (hyperbolic - space-time) and anti-symmetric (Euclidean space). Since anti-symmetric matrices normally have zeros on the leading diagonal, this might be a little confusing.

$$\text{Symmetric} \rightarrow \begin{bmatrix} \cosh\chi & \sinh\chi \\ \sinh\chi & \cosh\chi \end{bmatrix}$$
$$\text{Anti-symmetric} \rightarrow \begin{bmatrix} \cos\theta & \sin\theta \\ -\sin\theta & \cos\theta \end{bmatrix} \tag{22.1}$$

The first type is described by a rotation matrix that is symmetric in the variable, χ, and the second type is described by a rotation matrix that is anti-symmetric in the angle, θ. Note that $\cos(-\theta) = \cos\theta$.

We have seen that one of the defining properties of groups is multiplicative closure – any two permutation matrices in the same finite group always multiply together to form a permutation matrix in that same finite group. The other defining properties of groups are:

i. *Associativity:* All matrices are associative.

ii. *An identity:* The unit matrix (with one's on the leading diagonal and zero's everywhere else). In the 2-dimensional rotation matrices, the identity is when the angle is zero.

iii. *A complete set of multiplicative inverses:* Rotation matrices in general have all these properties.

$$\begin{bmatrix} \cosh 0 & \sinh 0 \\ \sinh 0 & \cosh 0 \end{bmatrix} = \begin{bmatrix} 1 & 0 \\ 0 & 1 \end{bmatrix}$$

$$\begin{bmatrix} \cosh \chi & \sinh \chi \\ \sinh \chi & \cosh \chi \end{bmatrix} \begin{bmatrix} \cosh \varphi & \sinh \varphi \\ \sinh \varphi & \cosh \varphi \end{bmatrix} = \begin{bmatrix} \cosh(\chi + \varphi) & \sinh(\chi + \varphi) \\ \sinh(\chi + \varphi) & \cosh(\chi + \varphi) \end{bmatrix}$$

Thus any particular rotation matrix is a particular group, but they each have an infinite number of elements – $\{0, \chi\}$ can take any real value. The above rotation matrices are groups with an infinite number of elements. Such groups are called Lie groups. Lie groups are concerned with 2-dimensional rotations in different types of space. Lie groups are expressed as the commutation relations between matrices that each represent one of the 2-dimensional rotations. The Lorentz group is the Lie group known as $SO(3, 1)$. In the Lorentz group, there are three symmetric (space-time) 2-dimensional rotations $\{(t, x), (t, y), (t, z)\}$ and three anti-symmetric (Euclidean space) 2-dimensional rotations $\{(x, y), (x, z), (y, z)\}$. Thus, the Lorentz group is six 2-dimensional groups of infinite order fitted together somehow to make a 4-dimensional space-time. As such, it has six identities and ought to be 12-dimensional, which makes it a bit of a "dog's dinner" group wise, but this is not different from the usual three rotation matrices that we associate with $\mathbb{R}^3$. We continue.

22.3 THE STANDARD PRESENTATION OF THE LORENTZ GROUP, $SO(3, 1)$, IN $\mathbb{R}^4$ SPACE

In the above name, the S stands for special, the O stands for orthogonal (as opposed to unitary or simplistic) which means that the rotations are in $\mathbb{R}^n$ space (as opposed to $\mathbb{C}^n$ or $\mathbb{H}^n$ space), the (22.1) stands for three space dimensions and one time dimension. Actually, the (22.1) stands for the distance function $d^2 = t^2 - x^2 - y^2 - z^2$, but that is seen as three space dimensions and one time dimension. Thus, $SO(3, 1)$ is the Lie rotation group of $\mathbb{R}^4$ space with distance function $d^2 = t^2 - x^2 - y^2 - z^2$.

Within our 4-dimensional space-time there are three spatial rotations (anti-symmetric), which we call rotations, and three temporal rotations (symmetric), which are changes of velocity and which we call boosts. It is conventional to take the view that a large rotation is an infinite number of infinitesimally small rotations, and, based on this, we seek the infinitesimal rotation matrix.

The spatial rotations in space-time are given as 4×4 matrices, we then let the angle, θ, approach zero. As $\theta \to 0$, $\cos\theta \to 1$ & $\sin\theta \to \theta$, (we can view this as grabbing the first term of the series expansions of the trigonometric functions and ignoring the higher powered terms) and so:

$$\begin{bmatrix} 1 & 0 & 0 & 0 \\ 0 & 1 & 0 & 0 \\ 0 & 0 & \cos\theta & -\sin\theta \\ 0 & 0 & \sin\theta & \cos\theta \end{bmatrix} \to \begin{bmatrix} 1 & 0 & 0 & 0 \\ 0 & 1 & 0 & 0 \\ 0 & 0 & 1 & 0 \\ 0 & 0 & 0 & 1 \end{bmatrix} + \begin{bmatrix} 0 & 0 & 0 & 0 \\ 0 & 0 & 0 & 0 \\ 0 & 0 & 0 & -\theta \\ 0 & 0 & \theta & 0 \end{bmatrix} \quad (22.2)$$

We put the identity to one side, multiply the angle by $i = \sqrt{-1}$ (see below), and normalize (divide by θ) to get:

$$J_1 = \begin{bmatrix} 0 & 0 & 0 & 0 \\ 0 & 0 & 0 & 0 \\ 0 & 0 & 0 & -i \\ 0 & 0 & i & 0 \end{bmatrix} \quad (22.3)$$

Which we call a generator of a spatial rotation in one spatial plane; notice that the matrix is anti-symmetric. If the reader thinks this is a little contrived, she is not alone. However, can the reader produce a better way of representing the spatial rotation of 4-dimensional space-time?

Starting with:

$$\begin{bmatrix} 1 & 0 & 0 & 0 \\ 0 & \cos\theta & 0 & \sin\theta \\ 0 & 0 & 1 & 0 \\ 0 & -\sin\theta & 0 & \cos\theta \end{bmatrix} \quad \& \quad \begin{bmatrix} 1 & 0 & 0 & 0 \\ 0 & \cos\theta & -\sin\theta & 0 \\ 0 & \sin\theta & \cos\theta & 0 \\ 0 & 0 & 0 & 1 \end{bmatrix} \quad (22.4)$$

Leads to:

$$J_2 = \begin{bmatrix} 0 & 0 & 0 & 0 \\ 0 & 0 & 0 & i \\ 0 & 0 & 0 & 0 \\ 0 & -i & 0 & 0 \end{bmatrix} \qquad J_3 = \begin{bmatrix} 0 & 0 & 0 & 0 \\ 0 & 0 & -i & 0 \\ 0 & i & 0 & 0 \\ 0 & 0 & 0 & 0 \end{bmatrix} \qquad (22.5)$$

Similarly, but with no i, we get the boosts (changes of velocity - temporal rotations):

$$\begin{bmatrix} \cosh \chi & \sinh \chi & 0 & 0 \\ \sinh \chi & \cosh \chi & 0 & 0 \\ 0 & 0 & 0 & 0 \\ 0 & 0 & 0 & 0 \end{bmatrix} \rightarrow K_1 = \begin{bmatrix} 0 & 1 & 0 & 0 \\ 1 & 0 & 0 & 0 \\ 0 & 0 & 0 & 0 \\ 0 & 0 & 0 & 0 \end{bmatrix}$$

$$(22.6)$$

$$K_2 = \begin{bmatrix} 0 & 0 & 1 & 0 \\ 0 & 0 & 0 & 0 \\ 1 & 0 & 0 & 0 \\ 0 & 0 & 0 & 0 \end{bmatrix}, \qquad K_3 = \begin{bmatrix} 0 & 0 & 0 & 1 \\ 0 & 0 & 0 & 0 \\ 0 & 0 & 0 & 0 \\ 1 & 0 & 0 & 0 \end{bmatrix}$$

The commutation relations between the above six matrices, $\{J_i, K_i\}$ are the Lie algebra of the Lorentz group. The use of i is connected to the distance function of space-time being:

$$\text{distance}^2 = t^2 + (ix)^2 + (iy)^2 + (iz)^2 = t^2 - x^2 - y^2 - z^2 \quad (22.7)$$

Note that the K_i matrices are the symmetric matrices and the J_i matrices are the anti-symmetric matrices.

There are commutation relations between these "rotation generating" matrices (which are usually called just generators). For example:

$$[J_1, J_2] = J_1 J_2 - J_2 J_1 = iJ_3 \qquad (22.8)$$

The commutation relations are:

$$[J_1, J_2] = iJ_3, \quad [J_1, J_3] = -iJ_2, \quad [J_2, J_3] = iJ_1$$
$$[J_2, J_1] = -iJ_3, \quad [J_3, J_1] = iJ_2, \quad [J_3, J_2] = -iJ_1, \qquad (22.9)$$

$$[K_1, K_2] = iJ_3 \qquad [K_1, K_3] = -iJ_2 \qquad [K_2, K_3] = iJ_1 \quad (22.10)$$

$$[J_1, K_1] = [J_2, K_2] = [J_3, K_3] = 0$$

$$[J_1, K_2] = iK_3 \qquad [J_1, K_3] = -iK_2 \qquad [J_2, K_3] = iK_1 \qquad (22.11)$$

$$[J_2, K_1] = -iK_3 \qquad [J_3, K_1] = iK_2 \qquad [J_3, K_2] = -iK_1$$

Notice the J_i in the second set of these. We have it that the commutator of two Lorentz boosts (temporal rotations in space-time) is a spatial rotation. To put it a different way: the commutator of two symmetric rotations is an anti-symmetric rotation – we have that in the A_3 algebras.

Aside: The Thomas precession is precession of spin in a magnetic field. In the rest frame, the spin polarization for a point electron obeys the equation:

$$\frac{d}{dt}\vec{S} = \frac{e}{mc}\vec{S} \times \vec{B} \qquad (22.12)$$

Above, we have mentioned that the Aharonov-Bohm effect shows that the potential of a field is a real physical thing. The equation above shows no precession of the electron polarization if the magnetic field is zero. In the Aharonov-Bohm experiment, there is no magnetic field, and so no precession of the electron polarization, but there is a potential, A_0, that corresponds to the phase, in the Euclidean complex plane, of the electron. The precession of this phase follows the equation:

$$\frac{d}{dt}F_k = iq\,A_0 \times F_k \qquad (22.13)$$

Where F_k is a Lie group generator.

The anti-commutators, $\{J_i, J_j\} = J_i J_j + J_j J_i$ play no part in the Lorentz group, and nor does the identity.

Another way of looking at the Lorentz group is that:

i   The commutators of the anti-symmetric matrices are anti-symmetric matrices.

ii. The commutators of the symmetric matrices are anti-symmetric matrices.

iii. The commutators of the anti-symmetric matrices and the symmetric matrices are symmetric matrices.

These are exactly the commutation relations that we have within the $C_2 \times C_2$ algebras.

The above commutation relations and generators are the Lie algebra $SO(3, 1)$.

22.4 SPLITTING THE LORENTZ GROUP

The standard procedure when dealing with the Lorentz group is to split the generators into two copies of the $SU(2)$ Lie algebra. This is of no great interest to us, but we include it for completeness. We form the combinations:

$$J_{i+} = \frac{1}{2}(J_i + K_i) \qquad \& \qquad J_{i-} = \frac{1}{2}(J_i - K_i) \qquad (22.19)$$

And we discover the commutation relations:

$$[J_{1+}, J_{2+}] = iJ_{3+} \qquad [J_{1+}, J_{3+}] = -iJ_{2+} \qquad [J_{2+}, J_{3+}] = iJ_{1+}$$
$$[J_{1-}, J_{2-}] = iJ_{3-} \qquad [J_{1-}, J_{3-}] = -iJ_{2-} \qquad [J_{2-}, J_{3-}] = iJ_{1-} \qquad (22.15)$$
$$[J_{1+}, J_{2-}] = [J_{1+}, J_{3-}] = [J_{2+}, J_{3-}] = 0$$
$$[J_{1+}, J_{1-}] = [J_{2+}, J_{2-}] = [J_{3+}, J_{3-}] = 0$$

Which is a mathematician's way of saying that we have two separate $SU(2)$ algebras, the J_i+ algebra and the J_{i-} algebra. This is written as $SU(2) \otimes SU(2)$. $SU(2)$ is the rotation group in the space formed from two copies of the complex plane being fitted together, $\mathbb{C}^2$. We say that $SO(3, 1) \cong SU(2) \times SU(2)$; the cross does not mean the same as it does in finite group theory or in the vector cross product but is purely notational.

The astute reader will have recalled that $SU(2)$ is isomorphic to the quaternion rotation matrix, and so we have broken the Lorentz group of six rotations in space-time into two sets of three rotations each of which is isomorphic to quaternion space. What does this mean physically? Looking at the above J_2+ matrix, we see it is not quite the same as the j part in the quaternion matrix. The non-zero elements occupy the corresponding positions, but they are not all the proper values.

22.5   A GENERAL VIEW

The standard view of space and time before Einstein produced the theory of general relativity was that space and time were a background within which the physics of universe plays out. It has often been pictured that space-time is a theater stage and that particles of matter and force are theatrical actors upon that stage. The theory of general relativity brings with it the idea that space interacts with matter and energy – mass-energy curves space (or at least makes it non-commutative). However, general relativity still sees space-time as very much a separate thing from the physics of the universe, and the connection is gravitational only. In earlier chapters, we have seen that electromagnetism (the Maxwell equations) is within the quaternions and thus within the finite group $C_2 \times C_2$. Also in earlier chapters, we might have found the 4-dimensional space-time that we observe, as expressed in the Lorentz group, within the finite group $C_2 \times C_2$. Thus, it seems that electromagnetism and space-time are just two aspects of the same finite group. Thus, in a sense, space-time might be an electromagnetic phenomenon like electrons or magnetic dipoles. This is a radically different point of view from the "space-time is the stage of the theater" point of view, even if we allow the stage to bend with the weight of the actors.

Aside: The standard model of particle physics associates each generator of the unitary Lie groups $\{U(1), SU(2), SU(3)\}$ with a boson (force particle). If we do the almost the same, and we associate each non-commutative generator with a boson and each commutative generator with a fermion (the other type of particle) and we use the algebras of C_2, $C_2 \times C_2$, $C_2 \times C_2 \times C_2$, then we get a match with the particle content of the universe excepting the Higgs boson. If we continue and include the C_2, $C_2 \times C_2 \times C_2$ group, we get, subject to the above assumption, super-symmetric unification. It might be just numerology.

Within the above finite groups, C_2, $C_2 \times C_2$, we find almost exact copies of the Lie groups $\{U(1), SU(2)\}$. Within $C_2 \times C_2 \times C_2$, we find an almost exact copy of the Lie group $SU(3)$. $C_2 \times C_2 \times C_2 \times C_2$ contains nothing like the Lie group $SU(4)$.

23

THE EXPANDING UNIVERSE AND THE COSMIC BACKGROUND RADIATION

In 1912, Vesto Melvin Slipher (1875–1969) discovered that the spectral lines from "galactic like nebulae" were red-shifted. At this time, it was generally thought that these "galactic like nebulae" were inside of, and part of, our galaxy, the Milky Way[1]. Such red-shift is an indication that those "galactic like nebulae" are moving rapidly away from the Earth. It was 1923 when Edwin Powell Hubble (1889–1953) measured the distance from Earth to those "galactic like nebulae" using cephcid variable stars and discovered that the "galactic like nebulae" were far too distant to be part of the Milky Way. Each galactic like nebula therefore had to be an independent galaxy outside of the Milky Way. By 1929, Edwin Hubble, together with Milton Lassell Humanson (1891–1972), had realized that there was a correlation between the distance of each independent galaxy from Earth and the amount of red-shift associated with that independent galaxy. This correlation is known as Hubble's law, and it says

---

[1]. There were individuals who thought differently. In the early part of the 1800s Herschel has expressed the opinion that the nebulae were outside of the Milky way. Similar opinion had been voiced by Kant, Swedenborg, Lambert, and Wright.

that the more distant a galaxy is from Earth, the faster it is receding from Earth (with appropriate numbers). The rate of recession against distance is independent of the direction in which we observe – it is isotropic. Later astronomy with much more capable telescopes (the Hubble telescope for example) and different types of telescope (radio telescopes etc..) have confirmed Hubble's findings, subject only to the fact that galaxies come in small gravitationally bound groups and it is these groups that are receding from each other.

We now have the view that there are some 10^{11} separate groups of gravitationally bound galaxies within our observation – each with some 10^{11} stars. With the exception of a few local galaxies that are bound to the Milky Way by gravitation, every one of these other galaxy groups is receding from us and the more distant a galaxy group is from us, the faster it is receding. This understanding is known as the expanding universe.

It seems that special relativity is not enough to explain the physical universe between galaxy groups. High frequency photons of light have more energy than photons with a low frequency. Thus red-shifted light has less energy when it reaches us than it did have when it left its source. We take conservation of (mass)-energy to be an inviolable law of physics, but it seems that it does not apply outside galaxy groups, or perhaps the energy is being converted into empty space. Since the universe seems to be infinite and the further a galaxy group is from us the faster it recedes from us, the very distant galaxies will be receding from us at superluminal velocity – faster than the speed of light - that's not allowed by the normal laws of physics.

At first thought, it is as if a giant explosion occurred in the distant past, like an exploding grenade, and each galaxy is a fragment of the matter in that explosion that has been flung out into empty space with the fastest moving now being the most distant from the point of the explosion. If this were so, then, unless the Earth was left at the very center of that explosion, space would seem to us to contain different numbers of galaxies in different directions. If the universe was the aftermath of a conventional type of explosion into already existing empty space, then, as we looked back towards the point of that explosion, we would see more galaxies, and a higher concentration of galaxies, in that direction than we see when we look away from that point of the explosion – picture yourself as riding on a fragment of an exploding grenade. This does not accord with observation – the universe is isotropic, and we do not like to think that the Earth is at the very center of the universe.

There are other difficulties with this conventional explosion scenario. The galaxies that are very distant from us must have been traveling very quickly away from us for a very long time. The processes of the universe slow down in rapidly moving galaxies as seen by a stationary observer. Thus, it should appear to us, having accounted for the time it takes light to travel from distant galaxies to Earth, that the stars in these very distant galaxies are much younger than the stars in nearby galaxies. Indeed, the very distant galaxies ought not to be there because the process of galaxy formation should have slowed by time dilation to the extent that the galaxies have not had time enough to form. We do not observe this slowing down of the processes of the universe in distant galaxies. Time dilation does not happen with distant galaxies that are moving very rapidly away from Earth – surely this violates the theory of special relativity.

From special relativity, we would expect the mass of the distant galaxies moving very rapidly away from Earth to increase. Surely, this ought to have consequences that we could detect, but we see no such effects.

Because our observations do not accord with what we would expect if the expanding universe were the result of a giant explosion into empty space, we do not accept that explanation. This leaves us rather stumped. We are compelled to take the view that the empty space between galaxies is stretching. The more space, that is the more distance between us and a distant object, the more this empty space stretches, and the faster the distant object appears to be receding from us. The distant galaxies have been likened to dots on a balloon that is being inflated[2]. Each dot (galaxy) recedes from each other dot (galaxy), and the more distant that dots (galaxies) are from each other, the faster they recede from each other. The view is that empty space is expanding – getting emptier and more extensive!

---

[2] The expansion of the universe, Arthur Eddington, 1931

It is a view with which no-one feels comfortable, but no-one can think of any other explanation that is acceptable. There is a "tired light"[3] explanation which supposes that light loses energy as it travels through space, by very small amounts, and this explains the red-shift observed from distant galaxies, but this has not found acceptance. The tired light idea must assume and explain a static universe and it proposes no explanation for the cosmic background radiation that we will meet shortly. There is also a "shrinking atom" theory which proposes that the charge of electrons was less in the past. If the electron charge had been less, atoms would have been larger and so the emitted spectral wavelengths would have been longer. Unfortunately, such a change in electron charge would have affected the fine structure constant in such a way that there would be no long lived luminous stars like the sun. Humankind does not understand expanding space – I think that includes everyone of us!

Perhaps, as Ernst Mach might have preferred, each gravitationally bound group of galaxies has its own local zero rate of rotation against which the rotation of a planet within the group could be measured and, similarly, its own local zero acceleration against which acceleration could be measured. Perhaps, if these spatial notions are determined within the local galaxy group, then all spatial concepts are determined only within the local galaxy group. As such, the recession of distant galaxies from each other would be meaningless. Thus, special relativity applies within each galaxy group but not to the space between each galaxy group.

If empty space is expanding (stretching), then, at some time in the past, all the empty space between galaxies would have been "unexpanded". It seems that, at some time in the past, all empty space would be of zero extent and there would be no space between galaxies. Following special relativity, we include time as part of space and take the view that space-time is expanding and that it started from zero extent. Observations lead us to put the zero extent of time to be at 13.8 billion years ago. However, there is no projection on to the time axis by the cosh( ) function that is less than unity.

With disregard to the cosh( ) function, the current view is that the universe began with a huge explosion of space and time, called

---

[3.] The tired light idea was proposed by Fritz Zwicky in 1929.

the big bang, 13.8 billion years ago, and that the matter in that explosion did not shoot out into empty space because there was no empty space then. It was at the big bang that space-time started and matter, riding on expanding space-time, was scattered throughout the universe. It seems that it is space and time that exploded and not matter. The reader can perhaps see why people who do not understand science think cosmology is no more than a different religion. It gets worse; in order to explain all the observations, we invent a thing called inflation. The idea of inflation is that, at the very start of the universe, space expanded extremely rapidly, indeed, much faster than the speed of light. It does explain the observation that the universe is very flat and uniform at all scales, but the proposed mechanism to drive this inflation (a scalar field with negative pressure resulting from a "super-cooled" phase transition) is unobservable and thus this mechanism is no more than an idea. The inflation scenario works only if a particular parameter of the theory is very finely tuned[4].

We have no understanding of stretching space-time, and we have even less understanding of inflating space-time, and we do not know what matter really is either; but we do know why pubs exist; they exist for cosmologists to drown their sorrows.

The mystery deepens. It used to be thought that the gravitational attraction between galaxies would slow the expansion of the universe. If the universe were like a grenade exploding into already existing empty space-time, then gravity would slow the expansion of the universe. It used to be thought that, eventually, 160 billion years from now, the universe would stop expanding and might start, under gravitational influence, to contract. Recent observations seem to indicate that the rate of spatial expansion is not slowing but is accelerating; very distant galaxy groups are receding from us faster than they ought to be. That ought to be spatial and temporal expansion acceleration did it not? but the observations show only spatial acceleration. It is as if some anti-gravity like force is accelerating distant galaxies away from Earth[5].

---

4. Roger Penrose has pointed out that a flat universe with no inflation is 10^{100} times more likely than a flat one with inflation.

5. Perhaps everyone just wants to get away from people who do not understand

We observe that the universe is spatially homogeneous. Since we live in space-time, we would expect the universe to be temporarily homogeneous, the same now as it was ten billion years ago. Observations of quasar counts and other empirical evidence indicate that the universe is evolving in time and is not temporally homogeneous. Of course, if it were that the universe was always 13.8 billion years old (including 50 billion years ago), with all the evidence to show this, then we would have temporal homogeneity, but that is not a commonly held view. It seems, the universe is not temporally homogeneous in an historical sense. We do, however, believe that the laws of physics, including the values of the physical constants, are the same now as they were in the past and will be in the future; time is homogeneous in that sense.

We conclude this section of the book with a little speculation. Let us consider the displacement vector of the hyperbolic complex numbers; these are the numbers that correspond to space-time. Displacement corresponds to age and to spatial extent. If we draw a diagram of these numbers on a sheet of (Euclidean) paper, they appear as existing only between the +45° line and the −45° line with age on the horizontal axis.

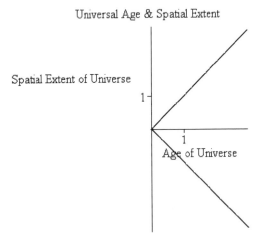

Universal Age & Spatial Extent

Spatial Extent of Universe

Age of Universe

---

empty space. In "The Expanding Universe" (1933), Eddington wrote, " The unanimity with which the galaxies are running away looks almost as though they had a pointed aversion to us."

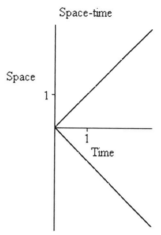

As we move positively along the age axis, we notice that the vertical distance from the +45° line to the −45° line increases. Is this the expansion of space with increasing age? Now, return to the velocity vector (space and time on the axes); the asymptotes are limiting lines that are approached by the ratio of the cosh( ) function to the sinh( ) function. Because the cosh( ) function is unity when the sinh( ) function is zero, the asymptotes are not the limiting velocity except at large values of the space-time angle. Suppose that limiting velocity is a hyperbola passing through unity on the horizontal axis. This means that the velocity of light was infinite at the start of the universe. Does this do away with the need for inflation at the start of the universe? As the hyperbola approaches the asymptotes, its slope, which is the velocity of light, varies from infinite to the limiting velocity of one. At some time in the past, the velocity of light was greater than one, and we would expect to see this in the form of distant galaxies accelerating in their recession from us, which we do.

23.1   THE COSMIC BACKGROUND RADIATION

In 1964, Arno Penzias (1933–) and Robert Woodrow Wilson (1936–), while working on ultra-sensitive micro-wave receivers for Bell labs, discovered a background radiation that came from all over the universe[6]. It is today called the cosmic background radiation (CBR). It had been predicted to exist by Alpher and Herman in 1948 based on Gamow's theory of element production in the big bang. The CBR was interpreted by Robert Dicke as the expansive remnant of a $3000K$ soup of protons and electrons in the big bang[7]. When the temperature dropped below $3000K$, the electrons and protons could combine to form atoms and the universe became transparent to radiation. The cosmic background radiation, CBR, accords with a background temperature of the universe today of approximately $2.725K$. The CBR has a typical wavelength of one millimeter and is about 400 photons per cubic centimeter.

The CBR is almost perfectly isotropic in its distribution across the whole of space, but there is a small dipole anisotropy corresponding to a temperature difference of $5 \times 10^{-3} K$ from one side of the universe to the opposite side[8]. Measured against the CBR, the dipole anisotropy corresponds to the Earth moving through space at 627±22 kilometers per second towards the constellation of Leo.

The CBR provides a reference frame for the whole universe. Special relativity says there is no absolute reference frame. So the CBR cannot be an absolute reference frame. The whole of physics works just the same when measured against reference frames other than that of the CBR. However, since the CBR is everywhere, the CBR is a universal reference frame against which the velocity of all things can be measured. As such, it provides a universal simultaneity and a universal time. It is measured against the CBR universal time that we say the universe has an age of 13.8 billion years. Every galaxy group sets its clock by the CBR, and so every galaxy group is the same age as all other galaxy groups. Of course, every galaxy group thinks that it is stationary and that it travels through only time at the speed of light.

Imagine identical twins, Alice and Zara, who were separated at the birth of the universe and who have been moving relative to each other ever since. The first twin, Alice, might be 13.8 billion years old

---

[6] Astrophysical Journal Vol. 142 pg. 419.

[7] Astrophysical Journal Vol. 142.

[8] This anisotropy was discovered by the Cosmic Background Explorer satellite, COBE, launched in 1988.

and think that, since she was born at the same time as the universe, the universe is 13.8 billion years old, but she will think her twin sister, Zara, to be only 2 billion years old. Since Zara has lived since the birth of the universe and is only two billion years old, the universe must be only two billion years old by Alice's reckoning. The same is true in reverse for Zara who will think herself to be 13.8 billion years old. Thus, each twin gives two different ages to the universe. It is only when they use the same reference frame, which the CBR provides, that the twins have only one age for the universe.

From the point of view of the stationary observer, light does not travel through time (it travels through only space), and so the stationary observer is of the opinion that light thinks the universe is still zero years old. However, from the point of view of a photon of light, light does travel through time (it is stationary in its own reference frame), and so light thinks the universe is 13.8 billion years old years old.

The CBR does not give us an absolute simultaneity or an absolute time (there is no absolute age to the universe, but there is a universal one). The universal time fits with the observation that the relative abundance of U^{235} to U^{238} indicates that stars have been burning for 1.5×10^{10} years. Since stars convert hydrogen into helium, we can also measure the relative abundance of hydrogen to helium and use this as a universal clock. This too corresponds to the universal time given by the CBR. Without this universal clock, stars would evolve at different rates in differently moving parts of the universe and space would not be homogeneous. We need a universal clock to keep the universe spatially homogeneous.

Thus, we see, that on a cosmological scale, we can largely ignore special relativity and work with the universal CBR reference frame.

Special relativity has nothing to say about "stretched" space-time, or "inflated" space-time, other than to remind us that we understand very little about these things. We note that our attempt to derive the finite group equivalent of the Lorentz group in the previous chapter also said nothing about "stretched" space-time, or "inflated" space-time.

CHAPTER 24

CONCLUDING REMARKS

I hope this book has given the reader a comprehensive understanding of the theory of special relativity. That is certainly one of the author's intentions.

Another of your author's intentions is to open up the nature of empty space for examination, and I hope that this book might have at least made a start in that endeavor.

I hope also that this book has given the reader an understanding of the geometric spaces inside the finite groups. This is an area of mathematics and physics that is still in its infancy. There are connections to particle physics within the finite group spaces that we have hardly touched upon and of which we know only very little at present. There is much to do and much promise of reward in this area. Perhaps the reader will go on to study in this field and contribute to the maturation of this area of human knowledge. Even if the reader does not so do, I hope that the discussion of the finite group spaces has enlightened the reader's understanding of the nature of space and time and convinced the reader that they need not accept the conventional view that space is no more than copies of the real numbers, $\mathbb{R}$, fitted together in some way.

We have tried to build our universe from no more than the C_2 finite group and its cross products. We might refer to our universe as the C_2 universe or the electromagnetic universe. We have hardly touched upon the C_3 finite group and not at all upon its cross products. If we can build a universe from C_2, then we ought to be able to build a universe from C_3. Almost nothing is known in this area. There are infinitely many other finite groups. Some of them are very different from the cyclic groups. Nothing at all is known about the geometric spaces in the lowest even order finite simple group A_5, but they is likely to be very different from the cyclic group spaces. Presumably, none of these other universes are detectable to we C_2 beings, or perhaps the C_2 sub-spaces of some of these will be detectable to us.

The usual modern view of the nature of scientific theories is that they are collections of observed facts that have been overlain by a systematizing understanding and that they approximate reality. The systematizing understanding derives from and is the invention of the human mind. Such a theory is a, severely constrained by observation, human-made amalgam of concepts and observational prejudices. The way we have tried in this book to build the theory of special relativity and electromagnetism from nothing more than numbers is a very different type of scientific theory. We develop our theory from no more than the observed existence of real numbers and finite groups; we then compare the result to reality. If the result fits our observations of reality, then our theory is more than just a systematizing understanding of observations – it is numbers. Numbers do not derive from the human mind. Nor have we relied upon mathematical axioms as statements of declared absolute truth. The nature of numbers that is captured in the axioms of algebraic fields is seen by this book as being the observable properties of numbers, like multiplicative closure or absence of zero-divisors.

We are not that different from Einstein. The theory of special relativity had its historical origins very much in mathematics rather than in observation, and this is even more true of the theory of general relativity, and perhaps even more true again of string theory.

We have come a long way from no more than real numbers and the finite groups. Considering that empty space is no more than nothing, we have derived a great deal from that nothing.

It is now time to return to the first page of this book and read it again. You are likely to gain a great deal more from this re-read than you gained from the first read.

BIBLIOGRAPHY

Armstrong, M. A.

Groups and Symmetry
ISBN 0-387-96675-7

Callahan, James

The Geometry of Space-time
ISBN 0-387-98641-3

Grant, I. S. &
Phillips, W. R.

Electromagnetism
ISBN 0-471-92712-0

Harrison, Edward

Cosmology
ISBN 0-521-66148-X

Lounesto, Pertti

Clifford Algebras and Spinors
ISBN 0-521-00551-5

Lucas, J. R. &
Hodgson, P. E.

Space-time & Electromagnetism
ISBN 0-19-85-2038-7

Rindler, Wolfgang

Relativity, Special, General, and
Cosmological
ISBN 0-19-850836-0

Ward, J. P.

Quaternions and Cayley Numbers
ISBN 0-7923-4513-4

Zee, A.

Quantum Field Theory in a Nutshell
ISBN 0-691-01019-6

INDEX